Materials Forming, Machining and Tribology

Series editor

J. Paulo Davim, Aveiro, Portugal

More information about this series at http://www.springer.com/series/11181

Hossam A. Kishawy · Ali Hosseini

Machining Difficult-to-Cut Materials

Basic Principles and Challenges

Springer

Hossam A. Kishawy
Machining Research Laboratory (MRL)
University of Ontario Institute of Technology
Oshawa, ON, Canada

Ali Hosseini
Machining Research Laboratory (MRL)
University of Ontario Institute of Technology
Oshawa, ON, Canada

ISSN 2195-0911 ISSN 2195-092X (electronic)
Materials Forming, Machining and Tribology
ISBN 978-3-030-07119-6 ISBN 978-3-319-95966-5 (eBook)
https://doi.org/10.1007/978-3-319-95966-5

This Springer imprint is published by the registered company Springer Nature Switzerland AG
The registered company address is: Gewerbestrasse 11, 6330 Cham, Switzerland

Preface

Machining or in the other word "material removal" or "cutting" is perhaps one of the oldest manufacturing operations and the most widely used one in almost all industries. The popularity of this process is inherently related to its versatility for use not only as a primary manufacturing operation to obtain different components with variety of shapes but also as a secondary operation after other manufacturing processes. As a secondary operation, the machining process is used to generate high-quality surfaces, which cannot be obtained by other manufacturing processes. Despite the advances in manufacturing technology over the past decades that made it possible to obtain high-quality surfaces using alternative manufacturing techniques, machining operations are still inevitable, in many cases, to achieve specific dimensional accuracy and surface characteristics.

Machining is accomplished when a hard tool, commonly known as the cutting tool, penetrates into a relatively softer workpiece. Because of the relative motion between the tool and workpiece, excessive material is removed in the form of chip. During the chip formation, energy is mainly generated in two main zones, namely primary and secondary shear zones. While the fundamental aspects of machining operations are almost similar, material properties greatly influence the mechanics of chip formation and heat generation, which in turn lead to different modes of tool wear. Therefore, to perform a successful machining process, appropriate process parameters such as cutting speed, feed rate, and depth of cut and appropriate cutting tool materials and coatings must be identified. These parameters are usually obtained through extensive machining and machinability studies. These studies have generated an extensive database of knowledge. The acquired knowledge enables toolmakers to come up with suitable tool materials and coatings capable of enduring the severe conditions that may be encountered during machining. Despite remarkable progresses and developments in resolving many machinability issues, metal cutting research is still ongoing and perhaps will be continued for many years to come.

Among many reasons for the ongoing need to perform machinability studies is the increasing demand for higher metal removal rates and the introduction of new materials. On the one hand, increasing the productivity drives the need for increasing the metal removal rate, which requires higher cutting speeds. This justifies the search for new cutting tools that can withstand the elevated temperatures generated at higher cutting speeds. On the other hand, the endless efforts of industry to find materials with superior characteristics to meet the design requirements and withstand extreme working conditions result in the development of new materials. Some of these new materials are obtained through alloying and/or heat treatments that improve the physical and chemical properties. In some other cases, the new materials are formed by adding hard constituents to the currently existing materials to improve specific characteristics such as wear resistance. The new materials have unique metallurgical characteristics and introduce new machining challenges, which make them difficult to cut. This further stimulates the research and studies to find appropriate cutting tools and proper process parameters.

Machining difficult-to-cut materials significantly increases the machining cost and dramatically challenges the economic feasibility of machining process. The difficulties encountered during the machining of these materials vary from one material to another. These difficulties can be related to the abrasive nature of the added hard constituents or the improved strength, toughness, corrosion, or temperatures resistance of the original material. Researchers and toolmakers are always searching for better tool materials and coatings as well as optimized tool geometries and edge preparation to overcome the ongoing challenges.

This book dedicates five chapters to the machining of different types of difficult-to-cut materials, namely hardened steels, titanium and its alloys, superalloys, composites, and ceramics. Each chapter starts with an introduction to the material followed by historical background, metallurgy and properties, and industrial applications. The discussion is then continued by a detailed description of challenges during machining, mechanics of chip formation, modes of tool wear, and appropriate cutting tool materials for machining of that particular material. The last chapter is dedicated to shed light on the environmental aspects of machining process and different approaches and strategies to reduce the environmental concerns. This includes different technologies for the application of cutting fluids and possible opportunities to minimize/eliminate the use of them during machining operations.

Oshawa, Canada
February 2018

Hossam A. Kishawy
Ali Hosseini

Contents

Chapter 1
Introduction

Abstract History of human being has always been closely tied to the materials they have had available. Ancient civilizations commenced their existence using the naturally available materials such as soil, stone, plants, and even bones. For a very long period of time during the history of mankind, humans survived using only these naturally occurring and easily obtained materials. Hence, searching for new materials with diverse range of characteristics and acquiring knowledge about their application have been a fundamental basis for human development and innovation since the early days. Challenged by the daily life, human has always been curious to discover new materials to achieve broader range of his ambitious desires. From this perspective, the appropriate material of choice was and still is determined not only based on availability, cost, efforts, and even the ease of implementation, but also based on the specific properties offered by the material to satisfy the part functionality and design requirements. These requirements include but are not limited to a broad spectrum of properties such as strength, toughness, heat resistance, corrosion resistance as well as essential tools, equipment, and manufacturing processes. Thus, materials and materials science have been the fundamental factors to the development of civilizations. The importance of materials to the development of civilizations is to the extent that the anthropologists classify the historical eras based on the materials used during that era such as the stone, copper, bronze, and iron.

1.1 Historical Background

1.1.1 Stone Age

The first materials used by humankind were primarily those either available in nature, e.g., stone and wood or those that could be acquired from animals such as skins, bones, fur, and feathers. These materials were mainly used to build weapons, shelters, and early forms of clothing to protect early humans from animal threats or severe weather conditions in the wild nature. This period of mankind history which is also known as Stone Age is believed to have begun about 2.5 million years ago

H. A. Kishawy and A. Hosseini, *Machining Difficult-to-Cut Materials*, Materials Forming, Machining and Tribology, https://doi.org/10.1007/978-3-319-95966-5_1

[1]. The required tools or weapons in the Stone Age were mainly produced by striking the stones together to accidentally form sharp edges. Although metals such as copper, silver, and gold, in their native state that is very soft, were also known by the ancient civilizations during the Stone Age, their applications were limited to decorative objects or ornaments. This was mainly due to the lack of knowledge in extracting the metals from their ores that confined the metal sources to those of purely available in nature.

In 8000 B.C., Stone Age man began to use copper due to its relative abundance in nature and also its appearance. Copper started to be widely used particularly after man learned how to melt and extract it from its ore, which is a mixture of copper and other minerals. However, evolution of copper at its early stages did not diminish the importance of stone, and man continued to use both of them for a long time. Coexistence of stone and copper in this period of history led to the so-called Copper-Stone Age [1]. Introduction of copper was a great step forward in history of material science. Copper had great formability and castability, which allowed the Stone Age man to give it any desired shape by either hammering or casting. Even though the copper was a remarkable discovery, Stone Age man faced several challenges in utilizing it. Due to the lack of knowledge and limited tools, gasses were trapped during casting, which made cast copper porous. The cast copper was also relatively soft and therefore could not be effectively utilized to build strong utensils. Therefore, the early human were encouraged to look for a stronger material to overcome the weakness of copper. This was the beginning of a new era in the history of humankind.

1.1.2 Bronze Age

A big progress in the history of materials and materials science took place when Stone Age man discovered that properties of copper could be significantly enhanced by adding certain materials. This important discovery probably happened by either accident or experimentation. The Stone Age man realized that some impurities from the copper ore could be transferred to the cast copper during solidification. Man also realized that these impurities could lead to changes in properties. Although by that time these materials may had not been individually recognized, nowadays they are known as arsenic, antimony, silver, lead, iron, bismuth, and tin. The amount of these impurities was not under control so as their effect on the mechanical properties. Man discovered that adding some materials in controlled amount greatly improve the material properties such as strength, hardness, and castability.

Analysis of the artifact found during the archaeological excavation in the Middle East revealed that arsenic was the first material employed by the Stone Age man as an additive to enhance the properties of copper [1]. However, evidences show that arsenic was only used for a limited period of time as its toxic nature and lethal side effects may have been realized. Tin was eventually found to be the perfect material to be used as an additive element to enhance the properties of copper. This was the evolution of an alloy known as bronze. The melting temperature of bronze (950 °C) was comparatively lower than that of copper (1084 °C) along with better fluidity to flow into the mold. Moreover, bronze was comparatively harder than copper immediately after casting, and its hardness could be even further improved by hammering. These characteristics made bronze an interesting material and enabled man to build stronger tool with lower porosity and ultimately better quality.

1.1.3 Iron Age

Historians and archaeologists believe that Iron Age commenced somewhere between 1500 and 1000 B.C. [1]. Iron was not a new material for early human, and meteorite iron had already been used as early as 4000 B.C., but the available resources of meteorite iron were limited and humankind was not able to use it widely. The Iron Age is referred to as a period of history generally occurring after the Bronze Age, marked by the widespread use of iron. However, extracting iron from its ore was still a challenging task even during the Iron Age since reaching temperatures even close to its melting point (1538 °C), using the available equipment at that time, was almost impossible. The only practical way was to acquire iron from slags formed during casting copper. Throughout the Copper-Stone Age and Bronze Age, man learned that iron oxide could be used as fluxing agent when casting copper. Iron oxide reacts with the unwanted particles during casting and form slag that can be simply separated. In attempt to find a more reliable source of iron than the meteorite iron and incapable of reaching temperatures high enough to melt the iron ore, man discovered that slag formed when casting copper can be used as a source of iron. It was determined that this slag contains iron in its reduced form which is nowadays known as bloom or sponge iron. Although it was great step forward, however, the struggle was yet far from being over. The iron content of the slag was porous and was not suitable for many applications. The man soon discovered that nearly pure iron in a compacted form could be obtained by repeated hammering of the slag at high temperature. The product of this process is what currently known as wrought iron. This approach helped humankind to produce iron by reducing its ore into sponge iron that required temperatures around 1000 °C. This was a feasible solution compared to directly melting the iron ore at 1538 °C, which was a temperature almost impossible to reach at that time.

Discovering the iron and establishing process to extract it from available resources was a great success for early man, but the pure iron obtained by the devised process was quite soft (softer than bronze) and very susceptible to corrosion when exposed to humid atmosphere. It is believed that around 1400 B.C., the man learned how to improve the properties of iron to achieve higher strength and better corrosion resistance. By that time, the man discovered that bloom or sponge iron could be significantly hardened by being exposed repeatedly to the cycles of heating in charcoal furnace up to 1200 °C followed by hammering. The heating cycle softens the bloom and hammering removes the slag and compacts it.

Moreover, during this process, which is currently referred to as heat treatment, the iron was repeatedly exposed to carbon monoxide (because of burning charcoal) which consequently diffused carbon into the iron surface. The final product of this process was an iron–carbon alloy which was significantly harder than bronze. Nowadays, this product is known as steel.

As mentioned earlier, early epochs of human civilization are typically named in terms of materials from which the weapons and tools were made. From thousands of years ago until now, the human being has been always willing to develop new materials to facilitate them in building new tools, equipment, and weapons to promote quality of life, defeat enemies, and ultimately gain supremacy and dictate policies.

Although materials science has been gradually evolved over different periods of human's life, it experienced a substantial growth during the nineteenth century and mostly twentieth century. This rapid progress has led to remarkable advancements through a broad range of materials, dominating those of previous eras. Majority of the materials that are currently being used in different aspects of our life have been developed in the past two centuries. Before 1800, the list of known materials was almost similar to that of 2000 years earlier.

1.2 Modern Engineering Materials

The development of steel and acquiring knowledge about modifying its properties was probably the most important event in the history of materials science that paved the way for the future progresses yet to come. In the early nineteenth century, science and engineering experienced a dramatic progress, which boosted the application of materials science into an advanced level. This progress was continued more rapidly, than ever before, in twentieth century. Series of inventions and innovations in transportation, oil and gas, power generation, automotive, and ultimately defense and aerospace industries necessitated the use of materials capable of carrying higher mechanical loads and surviving in corrosive environments while maintaining the weight and cost minimum. Moreover, these materials had to be capable of sustaining high temperatures that could soften or liquefy most of the conventional materials. The importance of comprising such functional characteristics cannot be overstated especially when it comes to oil and gas, power generation, defense and aerospace industries where safety and reliability are prime concerns.

The scientists and engineers were now facing new challenges they had to come up with a solution for. Since the available materials at late 19th and early 20th were not capable of sustaining such harsh service conditions, more advanced materials that had never been implemented in the past, at least in the large scale, would have to be introduced. The attempt to overcome the newly encountered challenges shifted the focus of material scientists and engineers to improving the properties of existing materials and developing new materials whose properties could be tailored for different purposes. The result of this attempt was introduction of new generation of materials and alloys comprising all or at least some of the following features:

- High strength-to-weight ratio
- High stiffness and toughness
- Heat capacity and thermal conductivity
- Hot handedness (maintaining their hardness and strength at high temperatures)
- Corrosion and oxidation resistance
- Low fatigue crack propagation rate (fatigue resistance).

These materials and alloys can be generally categorized under one of the following families:

- Steels
- Titanium and its alloys
- Superalloys (mainly include nickel, iron–nickel, and cobalt alloys)
- Metal matrix composites (MMCs)
- Ceramics.

1.2.1 Steels

As a widely used material, steel is a family of alloys that is obtained by mixing iron and other elements, mainly carbon. Steel is relatively inexpensive yet very popular in industry due to its high tensile strength, and it is mainly used in construction and structural engineering. Members of this family also have high modulus of elasticity, fracture toughness, and fatigue resistance. The mechanical properties of steel can be altered by changing the alloying elements as well as heat treatment. Members of this family are extensively used in industrial applications such as oil and gas and chemical processing (valves, pumps, pipes, and tanks), power generation (steam and gas turbine components, transmission shafts), automotive (bearings, gears, shafts, rods, transmission, suspension, and frame), construction (beams, plates, piles, and sheets).

1.2.2 Titanium and Its Alloys

Titanium and its alloys are interesting materials with high strength-to-weight ratio, great stiffness, toughness, and fatigue resistance. Retaining their mechanical properties at elevated temperatures along with excellent corrosion resistance makes them materials of choice for many applications in aerospace and defense industries where heavy loads, high temperatures, and corrosive environments are very common. Airframes and jet engine components are the examples of such applications.

Titanium and its alloys are also used in applications other than aerospace such as steam turbine blades, marine and offshore applications, and automotive. Due to their great biocompatibility, they are also being used in biomedical industry to make prosthetic joints and implants.

1.2.3 Superalloys

Superalloys are a group of heat-resistant alloys that are mainly acknowledged for the capability of maintaining their mechanical characteristics such as strength, stiffness, and toughness at very high temperatures. The materials belonged to this category also exhibit great degradation resistance in terms of oxidation and corrosion. Nickel-based alloys, iron–nickel alloy, and cobalt alloys are the members of this family. These materials are mainly used in applications when the components will be exposed to very high temperatures. Components of gas turbine compartments, turbine blades, and turbine disks are examples of such applications.

1.2.4 Metal Matrix Composites (MMCs)

A composite material is defined as a material made from two or more constituents with considerably different characteristics. The main constituents of any composite materials are referred to as matrix and reinforcement. Matrix is the main material that dictates the shape of composite body by surrounding the reinforcement and holding them in position while reinforcement improves the matrix properties by imparting their superior mechanical characteristics. In metal matrix composites, the matrix material is made of a metal, whereas the reinforcement can be metal or any other material that can promote the mechanical, physical, or chemical characteristics of the matrix metal. Metal matrix composites are attractive material as their properties can be tailored to achieve high strength-to-weight ratio, high stiffness, and good damage resistance. Owing to their lightweight and high stiffness, MMCs are mainly used in aerospace industries, aircraft structures, and missiles wings and fines.

1.2.5 Ceramics

Although finding a unique definition for ceramic materials is not easy, ceramics are usually defined as the combination of metallic and non-metallic atoms in different possible arrangements. Ceramics are well known for their low density, high hardness and wear resistance, hot hardness and resistance to chemical degradation. Therefore, ceramic materials are mainly implemented in heat exchangers, some engine parts, electric and electronic components, and bio-applications such as implants.

1.3 Superior Characteristics, Major Challenges

Development of advanced materials revolutionized the knowledge of humankind in materials science and hastened the progress rate of industry. However, despite all of their rewarding features, advanced materials introduced new issues and challenges that have never been confronted before. Some of these materials are very expensive and require labor-intensive process to be produced. Moreover, due to their superior characteristics such as high strength, stiffness, and hardness, manufacturing processes that can be easily employed for most of the engineering materials may not be efficient when being implemented on the advanced materials. One of the most widely used manufacturing operations in the industry is machining in which a wedge-shaped cutting tool removes the material layer by layer from the workpiece surface and yields the desired shape and geometry. Advanced materials such as titanium and its alloys, structural ceramics, superalloys, metal matrix composites and steels (especially in the hardened state) rapidly deteriorate the cutting tools and impose several issues to the quality of parts and economy of the process. This rapid deterioration is mainly caused by high temperatures and stresses generated during machining. Rapid tool wear due to the abrasive nature of some of these materials is also a challenge for the machining industry. Due to the above-mentioned reasons, these materials are notoriously difficult to machine; hence, they are usually being referred to as difficult-to-cut or hard-to-cut materials.

This book is intended to present the history of these materials, their mechanical and chemical properties, industrial applications, and ultimately the issues that may be raised during machining of them. Modes of tool wear and the most suitable cutting tools for machining these materials will also be discussed in this book.

Reference

1. Hummel RE. Understanding materials science: history, properties, applications. Springer Science & Business Media; 2004.

Chapter 2
Hardened Steels

Abstract Metals, specifically steels, have transformed from their limited use by early blacksmiths to the current state of industrial mass production. The gradual progression of steelmaking processes has led to advancements in manufacturing processes, the quality and performance of products, as well as improving economies. The current chapter covers the machining of hardened steels, which is also known as hard machining. Hard machining refers to the process whereof a cutting tool removes the material from the surface of a workpiece with hardness value over 45 HRC and it can reach even up to 70 HRC. Hard machining can be achieved by almost all of the conventionally used machining operations such as hard turning, hard milling, hard boring, and hard broaching. The main objective of this chapter is to present important information about hardened steels. It briefly provides general information about hardened steels, their history of evolution, and a description of their unique mechanical and metallurgical characteristics. More importantly, it discusses the problems associated with manufacturing hardened steel parts and possible ways to overcome their machining difficulties. In this chapter, the machining operations that can be utilized for machining hard materials are investigated with a main focus on the application of hard turning and hard milling. The main challenges in the machining of hard materials, particularly hardened steels, applicable tool materials, required machine tool specifications, and attainable surface integrity will also be reviewed.

2.1 Introduction

Hardened steels are ferrous alloys that contain low to high alloying elements and have undergone a heat treatment process to increase their hardness, strength, and/or wear resistance. These properties are the key characteristics of hardened steels. The base material for these alloys is iron, which is why steel alloys are also called ferrous alloys; ferrous refers to iron. Steel alloys contain varieties of other elements such as manganese, nickel, chromium, titanium, tungsten, and most commonly carbon. The main element that designates steels from other ferrous alloys is carbon.

H. A. Kishawy and A. Hosseini, *Machining Difficult-to-Cut Materials*, Materials Forming, Machining and Tribology, https://doi.org/10.1007/978-3-319-95966-5_2

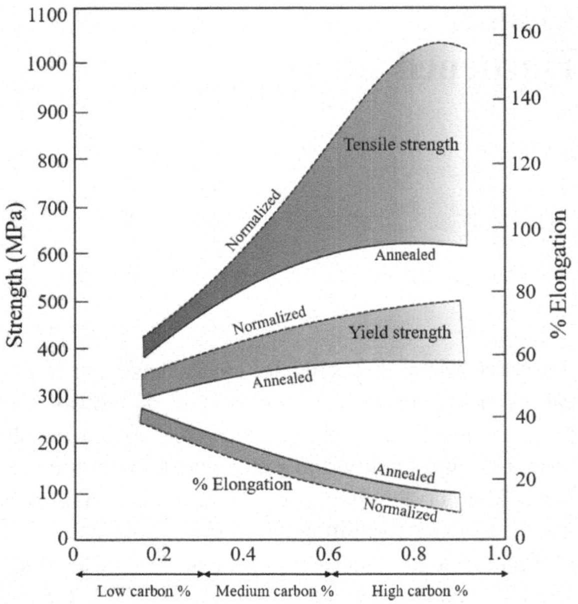

Fig. 2.1 Effect of carbon content on the mechanical properties of carbon steel [2]

According to the iron–iron carbide phase diagram, carbon can form a solid solution with iron up to maximum 6.70 wt%; however, from commercial point of view, steels are iron–carbon alloys for which the carbon content ranges approximately from 0.06 wt% to maximum of 2.0 wt% [1]. Generally, the more the carbon contents, the higher the yield and tensile strength. The effect of carbon content on the tensile and yield strength of steels is shown in Fig. 2.1. With the addition of other alloying elements, different material properties can be achieved and the unique characteristics of each alloy define its application. For instance, manganese is often used in steel alloys to enhance its ductility and improve its characteristics in metal forming applications. Typically hardened steels are entirely composed of iron except 0.15–2.0 wt% of carbon and smaller amounts of other alloying elements [2].

Several different treatment processes can be used to produce and enhance the properties of hardened steels. They include heat treatment, cryogenic treatment, case hardening (in terms of carburizing, gas nitriding or dissolution of nitrogen and hard nitride precipitations, induction hardening, flame hardening), and tempering.

2.1.1 Heat Treatment

The heat treatment process, also known as thermal treatment, involves a controlled sequence of heating and then cooling steel to particular temperatures that results in a material with improved strength and hardness [3]. This process is done with medium to high alloy content steel, anything above 0.30 wt% alloy. First, the temperature of steel will be increased to a high temperature that is below the temperature where the austenite microstructure is transformed to a pearlite structure. This is done for certain duration of time, and then, the steel is removed and quickly put into a quenching media where it is rapidly cooled. The goal here is to transform the austenite microstructure to a martensite microstructure. When the austenite is rapidly cooled, the face-centered cubic (FCC) structure of the austenite is transformed to a body-centered tetragonal (BCT) structure; forming the martensite phase [2]. Among the quenching media, water is one of the best ones when maximum hardness is desired; but where hardness can be sacrificed, oil may be used instead. Although water is a great quencher, the rapid cooling may lead to cracking or deformation resulting from the sample cooling unevenly. Quenching may also be done using gases where nitrogen is commonly used at approximately 20 bar. Helium is also a good option as it has a greater thermal capacity [4]. The quenching must be performed fast enough (high cooling rate) to reduce the likelihood of retained austenite. Retained austenite is a case in which only a portion of the austenite is transformed into martensite. Retained austenite results in dimensional instability, cracking, and reduction in hardness and strength. Therefore, the hardening process may need to undergo a few more cycles to get a good finished product. The martensite phase does not have as many slip systems as other structures. This makes martensite extremely hard and brittle and reduces its toughness. The brittleness and lack of toughness are generally undesirable in many applications; thus, the hardened material must then be tempered [2].

Tempering is a process that is done to balance the hardness with toughness. Needless to mention that toughness is the ability materials to plastically deform and thus absorb energy without fracturing. There are different ways to temper steel; a common method is martempering, which is a similar process to hardening and it involves heating and quenching of the steel. In martempering, the steel is heated to a temperature where ferrite is transformed to austenite; this temperature is called the austenitizing temperature. The heated steel is then put into another quenching media in the form of a hot fluid such as molten salt or oil. Once the steel has a uniform temperature throughout the part, it is removed from the quenching media and allowed to air cool, which makes the martensite within the material tempered [2]. After tempering, the steel becomes less brittle and becomes tougher, and has less hardness. Even though the hardness is reduced during tempering, the material is stronger, and more wear resistant than it was prior the heat treatment process.

2.1.2 Cryogenic Treatment

Cryogenic treatment is another method to harden the steels instead of the thermal treatment process. Like thermal treatment, this process is performed on the medium to highly alloyed steels. In cryogenic treatment, the temperature of the steel is lowered from room temperature to approximately −180 °C or even lower to −196 °C [5]. The steel will be held at this temperature for a period of 30 h [5]; however, the process period may vary between 24 and 36 h. It is important that the temperature is gradually lowered to avoid thermal shock and damages to the material. A temperature reduction rate of 2 °C per min is often used. When the steel is left at this low temperature, austenite transforms to martensite. This process is slow; however, it results in a near-complete conversion, which is better compared to the thermal hardening process, which results in 50–90% conversion. Once this process is complete, the steel still must be martempered in order to reduce the brittleness and increase the toughness [2].

2.1.3 *Case Hardening*

Case hardening, which is also referred to as surface hardening, is a process utilized to harden a metal surface, but maintains its core soft without any change. The objective of this technique is to increase wear resistance, thereby improving the surface quality of the material. Case-hardening treatment is widely employed in the automotive part industry. It is used to enhance the surface quality of gears and also used for a variety of steel forgings [6].

Unlike thermal treatments and cryogenic treatments, case hardening is done with low alloy steel, anywhere from 0.15 to 0.20 wt% carbon [3]. In addition, case hardening is a surface treatment, which results in a hardness gradient from the surface inwards with the inside of the steel still being tough and ductile. Therefore, there is no need to temper the steel afterward as in the case of thermal treatment and cryogenic treatments. In its simple form, the process involves heating the steel in an atmosphere that contains elements that alter the microstructure and composition of the steel surface, thus changing the surface properties. These elements can include carbon, nitrogen, and boron. The resulting material is a steel alloy with a surface that is hard, strong, and resistant to wear, but it gets softer, tougher, and more ductile beneath the outer surface.

2.1.3.1 Carburizing

Carburizing is a widely used method of case-hardening process in which steel with low carbon contents is heated (usually in the range of 850–950 °C) in presence of high carbon content material, such as charcoal. Under the effect of heat, carbon is diffused into the surface of the steel, forming a metal with greater hardness and strength. The low carbon content allows steel to be easily manipulated into different shapes and sizes before the carburizing process. Surface hardening is utilized to provide resistance to corrosion wear. In addition, products that undergo high pressures or impacts are generally case-hardened to withstand external shocks [7]. Products manufactured through this process have a relatively longer operational life span.

2.1.3.2 Gas Nitriding

Gas nitriding is another case-hardening process based on the reaction between nitrogen and steel at high temperatures. This hardening process does not require the involvement of quenching to achieve the final product. There are many similarities between the outcomes of gas nitriding and carburizing where both result in higher surface hardness and enhanced wear resistance. Prior to performing the process, austenitic stainless steels and ferritic steels must undergo annealing and stress relieving to minimize the size during the process. In the case of tight tolerances, surface of the workpiece can be polished by grinding to remove the outermost brittle layer formed during the process. Gas nitriding process can be used for the hardening of austenitic stainless steels only if they are annealed to ensure that the nitride case does not blister or flake. The hardened layer produced by gas nitriding, when implemented on austenitic stainless steels, is very thin; therefore, this form of steel hardening is only utilized for very specific applications such as the cases in which the steel must be wear resistant and also non-magnetic. Martensitic stainless steels hardened by gas nitriding offer high strength in the core that provides support for the surface [8]. Generally, in order to perform nitriding, special steel grades must be used to obtain high surface hardness. In comparison to a material that has undergone carburization and quenching, a steel processed with nitriding attains a superior surface hardness [6].

2.1.3.3 Induction Hardening

Similar to other case-hardening processes, induction hardening enhances wear resistance and increases surface hardness through the formation of a hardened surface layer. These enhanced surface characteristics are achieved while the material core remains unaffected. In this process, the metal is heated to extreme temperatures within a copper coil by means of an alternating current. The current flow induces an alternating magnetic field and thus generates heat, which in turn heats the outer surface of the metal. Subsequently, quenching must take place before obtaining the final material. The only form of steel that can be hardened using induction hardening is the martensitic stainless steel. Induction hardening is used for materials that experience heavy loading, specifically torsional loading, as well as for surfaces that endure severe impact loads. Symmetrical components such as gears and shafts that are used in automotive applications are usually hardened by means of induction hardening to reduce their wear.

2.1.3.4 Flame Hardening

The process of flame hardening is accomplished by heating the outer surface of steel parts using a flame formed by burning acetylene, propane, or natural gas. The heat created by the flame must reach extreme temperatures to harden the exposed surface. The process is then followed by quenching the heated steel in water. Flame hardening is advantageous especially when hardening flat surfaces such as flat wear plates is desired [9].

2.2 Historical Background and Evolution of Hardened Steels

Steel has evolved significantly over time; according to the World Steel Association, there are now more than 3500 different grades of steel and approximately 75% of modern steel was developed within the past 20 years. Hardened steel has historically been utilized for weapons, tools, and machinery parts [10, 11]. The earliest evidence of the invention of steel dates back to late thirteenth or early twelfth century BC [6]. Blacksmiths discovered that iron would become harder when left in a charcoal furnace. This was in fact the earliest form of case-hardening (carburization) treatment.

During the third century BC, Wootz steel was first made in India [12, 13]. It is a type of crucible steel portrayed by an arrangement of bands formed in high- or low-carbon steels. Steel production moving forward was mainly used for creating tools, weaponry, and armor. The early Romans were the first to discover a method of reducing brittleness in strain hardened steels for their combat equipment, which we now call tempering. There is evidence that tempering actually began in Palestine [14], where tempered martensite was found on a pick axe handle, dating back to 1200–1100 BC. In the third century AD, the Chinese learned how to manufacture high-quality steel, done through a process of melting iron and adding alloys like carbon to the molten iron. In the eleventh century AD, Damascus steel was made in the Middle East. This high-carbon steel was continually folded and forged to create high-strength swords. Due to the high alloy content and manufacturing process, these swords were very high in strength and hardness. Onward steel advancements had a slow but steady progression as steel products gained popularity. As production grew, more studies were done on steel's properties; in the sixteenth century, Vannoccio Biringuccio, an Italian metallurgist, published his work in what is known as the first systematic detailed text on metallurgy.

It was not until toward the end of the medieval times that adequate technical progresses were made in Europe with regard to steelmaking [4, 15]. The technical advances in the late eighteenth century made it possible to differentiate between iron and steel based on their carbon content [4, 15]. Mysticism and empirical experimentation were responsible for most of the primary understanding of quenching at the start of the Iron Age [4, 16, 17].

The formation of crucible steel occurred in 1751, and it was discovered by Benjamin Huntsman. He melted blister steel at extreme temperatures and used coke as the fuel in clay crucibles. The crucible process grew internationally, increasing the production of steel throughout places such as France, Tokyo, and Pittsburgh.

The benefit accrued from using the crucible process was the production of alloy steels [11]. Around 1850 AD, at the beginning of the industrial age, societies began to quantify and comprehend the heat treatment and quenching mechanism that had once been used to increase the effectiveness of knives, swords, and armor [4]. A major contributor to the increased demand for wrought iron was the industrial revolution, when humans started to use machines and equipment to manufacture products. However, wrought iron could not be mass-produced until the late eighteenth century.

Through the emerging need of industry for a more economically feasible process to produce steel during the eighteenth century, Henry Bessemer introduced his method in 1856. By implementing this method, impurities could be removed from impure iron to achieve an alloy with desired characteristics and amounts of different elements. Bessemer designed a receptacle, also known as a converter, where the iron would be heated in the presence of passing oxygen. Oxygen would react with the carbon content of iron and remove the carbon in the form of carbon dioxide. The Bessemer process also caused silicon to be eliminated from the iron ore. Passing the oxygen through the melted iron caused too much carbon to be lost, and the process had to be improved. To address this issue, Robert Mushet, a British metallurgist, started testing a compound of iron, carbon, and manganese that could remove oxygen from molten iron and add carbon. This compound was known as spiegeleisen [18]. Bessemer began adding spiegeleisen to his molten iron that solved the issue of the oxygen and the low amounts of carbon. However, one more problem still remained, Bessemer failed to remove phosphorus. The issue with phosphorus is that it causes the iron to be brittle, which means that it cannot be used in many industrial applications. A British inventor named Sidney Gilchrist Thomas found that the undesirable phosphorus would be successfully removed from the pig iron by adding limestone to the Bessemer process. This method revolutionized steel and made it a viable option for use in structures, as it was not brittle anymore and was strong to build upon structures.

In terms of hardening processes, boronizing was developed in 1895 by Henri Moissan and nitriding was patented in 1908 by Adolf Machlet [19]. These are popular case-hardening methods used to produce surface-hardened steels. A major and notable invention that helped the advancement in steel treatment technology was the electron microscope in 1931. With the use of an electron microscope, scientists could gain a much better understanding of material properties and structures before and after hardening processes. Due to this invention, much more refinement could be made in heat treatment processes that resulted in stronger and harder steels.

The manufacturing of steel was further enhanced by German engineer Karl Willhelm Siemens' creation of the open-hearth process. In the proposed process, excess carbon and other impurities would burn off using high temperatures. This method gave industry the capability to produce much larger quantities of steel at lower prices. Using the principle of heating a furnace, the industry was again revolutionized by the method known as electric arc furnace steelmaking. This method was introduced by Paul Heroult, and it was designed to pass an electric current through the iron. This would cause a reaction and would result in exothermic oxidation, and would result in high temperatures up to 1800°. By World War II, this method was being overwhelmingly used for the manufacturing of the majority of steel alloys. This method is also especially helpful because the operation can be stopped and started at any time with little to no financial loss. This method accounts for about 33% of global steel production. The next revolution that happened in steelmaking was oxygen steelmaking, which currently accounts for 66% of global steel production. Oxygen steelmaking is a process in which large quantities of oxygen is blown into molten iron. This is the most efficient method, and it has made open-hearth steel manufacturing unfeasible and is now not in use anymore. This method is much quicker than open-hearth methods. Some large vessels can convert iron into steel in less than an hour.

Another major step in steel hardening was in 1968 with the invention of new vacuum carburizing techniques. As another advancement in case hardening, this new invention greatly improved surface hardening processes. These advancements have all, in their own way, played crucial roles in steel development, especially in terms of hardened steels. They have brought us to the point we are at today, where we have extremely strong and hard steels, which can be used in a variety of equipment, such as tools, bearings, automobile mechanisms, and many other machines.

The above gives a brief summary on the history of steels and their evolution. The applications of steels in modern industry will be presented in this chapter after their mechanical and metallurgical properties are presented.

2.3 Metallurgy of Hardened Steels

As discussed in previous sections, hardened steel refers to a ferrous-carbon solution that has been heated to a high temperature and held until austenite phase is formed. It is then rapidly quenched into a hard, yet brittle martensite [20]. At this stage, the brittle steel can be tempered by reheating the unstable martensite, causing diffusion of the carbon atoms resulting in the formation of ferrite and cementite and then being cooled again [21]. This process results in a material that is less hard than the substance found after the initial rapid quenching, but much less brittle, making it a more desirable metal [20]. Tempering of martensite is dictated by the following reaction:

$$\text{Martensite (BCT, single phase)} \rightarrow \text{Tempered Martensite } (\alpha + Fe_3C \text{ Dual Phase})$$

The reaction reveals that the tempered martensite consists of small cementite particles that are evenly distributed within a continuous ferrite matrix. The cementite reinforces the ferrite. The continuous ferrite matrix has greater ductility and toughness in comparison to non-tempered martensite while still being nearly as

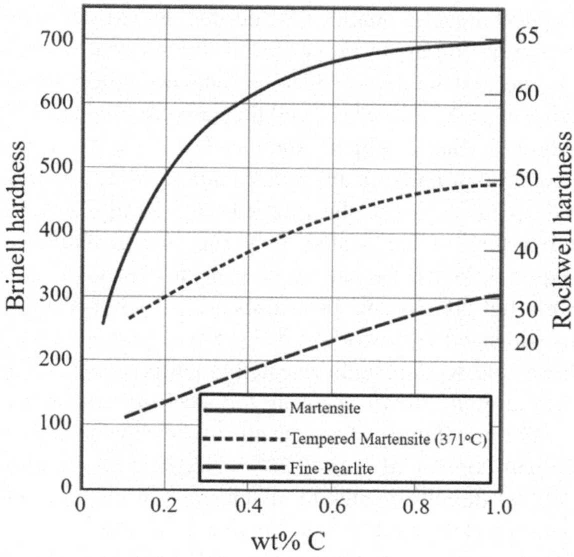

Fig. 2.2 Hardness as a function of carbon concentration for plain carbon martensitic, tempered martensitic, and pearlitic steels [20]

hard and strong [20]. The comparative brittleness between tempered martensite, martensite, and pearlite can be seen in Fig. 2.2 as a function of steel carbon content by weight percentage.

The microstructure of hardened steel and its corresponding mechanical properties vary based on certain parameters. The most influential parameters in hardening are the carbon content of the steel, the temperature to which the steel is heated before rapid quenching, and the temperature and time of quenching during the tempering of the martensite.

As stated earlier, the hardened steel is created when a steel specimen is heated to an extremely high temperature and then it is rapidly cooled using quenching media. The rapid cooling reduces crystallinity resulting in amorphous metal, which improves hardness. The important step to harden steel is called quenching. Quenching is necessary because cast steels and iron have a pearlite grain structure. This results when molten steel or iron is cooled very slowly. The pearlite is relatively soft and is not preferred for most applications of steel. When the solid solution above the transformation point is solid, a eutectoid transformation may occur. Raising the temperature of pearlite above its eutectoid point and then rapidly cooling it transforms the crystal structure into a much harder structure known as martensite. Martensitic structures are often preferred where high resistance to deformation is desired. This includes applications such as cutting edges of blades.

Hardened steels include a broad range of medium- to high-carbon steels, as the term hardened comes from the process where the steel has been tempered and quenched. When conferring the mechanical and metallurgical characteristics of hardened steels, it is beneficial to look at the different types of carbon steels and discuss their properties. These steels can be classified into two categories: rimmed steels and killed steels. Rimmed steels are great to use when pristine surface finish or good drawing qualities are needed. Killed steels are great to use when difficult to stamp or non-aging properties of the material is required.

The first family of steels is the low-carbon steels. These steels have a carbon content of 0.15% or less, and they are predominantly used where cold formability is required. Steel, which contains 0.05% carbon, is usually used for producing deep-drawn parts or any other parts needing severe deformation, hence it is considered as a part of the rimmed steel family. Low-carbon steels have low tensile properties; consequently, they are not generally used where high strength is required. When the carbon content increases, the strength of the steel increases as well, but unfavorable properties arise, such as decreasing in ductility or the ability to withstand cold working [22]. Steels that are classified as low-carbon, and have been cold-worked and/or heated to temperatures greater than 640 °C are inclined to drastic grain growth, causing the steel to become brittle.

The next family of steels can still be considered low-carbon steels, and range in a carbon content of 0.16–0.30%. Needless to mention that the increase in carbon content directly relates to an increase in strength and hardness but a decrease in ductility. In regard to heat treatment, the steels in this range of carbon content are also known as carburizing or case-hardening grades. In order for case hardening to occur, in addition to the carbon increase in the material, there must exist an increase in manganese as well. This is because when manganese is increased in the steel, the hardenability of the core and the case increase as well. Sometimes the design may require the use of rimmed or semi-killed steel, which falls under the family of low-carbon steels. These steels can find applications in many different sectors in industry. As previously described, any part that requires hardening at the surface uses steel of this grade. Other application of this grade of steel can be found in a vehicle transmission, rear axle gears, fan blades, some frame members, low-strength bolts, and some welded tubing.

The next group of steels is medium-carbon steels. These steels range in a carbon content of 0.30–0.52%. Different heat treatment processes and/or cold working are utilized to harden and strengthen this group of steels. Medium-carbon steels are classified as killed steels, and they are used in many automotive applications such as the construction of shift and brake levers. These steels have great forging capabilities; hence, they are used to make shifter forks and many other forgings where moderate strength is required. Higher carbon content is usually used to make crankshafts where high hardness after oil quenching is required [22]. The steels that belong to this family are also used in the form of wire and rods to produce bolts and studs.

The final family in hardened steels is the high-carbon grade. These steels range in a carbon content of 0.55 to 0.95%. These steels have a carbon content that is much greater than necessary for achieving maximum quenched hardness. The uses of high-carbon steels are found predominantly where wear characteristics need to be improved, such as with springs.

High-carbon steels are generally used in the spring industry where carbon content ranging anywhere from 0.65 to 0.85 can be used. High-carbon steels have also found their way to being used in the farm industry, as they have good wear properties when they have gone through appropriate heat treatment. Such applications in the farm industry require parts to be made of a carbon content of 0.70% up to 0.90%. Parts such as, plow beams, plowshares, rake teeth, scrapers, cultivator shovels, mower and binder sections, twine holders, and knotter disks are usually made from this grade.

2.4 Characteristics of Hardened Steels

Understanding the mechanics of machining of hardened steels without describing their inherent properties is incomplete. As a result, the mechanical properties and machining characteristics of hardened steels must be thoroughly investigated prior to performing any research in the field of hard machining. From the machining standpoint, hardened steels are recognized by their following characteristics [23, 24]:

2.4.1 *High Indentation Hardness*

Owing to their high hardness, the hardened steels are considered relatively abrasive and they can rapidly destroy the cutting tools. Therefore, the proper cutting tools for machining these materials must have high abrasion resistance [23, 24].

2.4.2 *Low Ductility (Brittleness)*

Hardened steels have low ductility and are very brittle. During metal removal, chips are formed under brittle fracture mechanisms rather than the continuous shear deformation [23, 24]. The formed chips during machining hardened steels are usually designated by unique morphology known as saw-toothed chips.

2.4.3 *High Hardness/E-modulus Ratio*

High hardness/E-modulus ratio improves the workpiece material ability for elastic recovery. However, this characteristic is one of the main sources of geometrical and dimensional errors when it comes to machining hardened steels.

2.4.4 Corrosion Sensitivity

Hardened steels are also affected by corrosion and wear. Iron naturally forms various forms of oxides. When exposed to the atmosphere, it reacts with oxygen and it becomes oxidized. Oxidation in iron is usually referred to as rust. The rate of corrosion depends on the atmosphere and the material exposed to it. Other factors that affect the rate of corrosion are temperature, oxygen, airborne salts, humidity, and other airborne chemical.

In regular atmospheric air, corrosion occurs mainly due to oxides and carbonates. In industrial atmospheres, rust can also be caused by sulfuric acid in the atmosphere; while near the ocean and bodies of water, it can be caused by water and salt. Acids also contribute heavily to the corrosion of steel. Sulfates and chlorides form a corrosive electrolyte on the surface of the steel. The most common electrolyte, which causes rapid rusting, is sulfur dioxide, SO_2, which is commonly found in the atmosphere of industrial environments. Sulfur dioxide is absorbed into the steel surface, but the absorption rate depends on the humidity and the moisture of the air. At concentration levels below 90%, sulfuric acid is corrosive to steel [14]. In general, contact should be avoided between hardened steels and dilute acids.

Corrosion to hardened steels can be avoided by several methods. The first method is to design the specimen in a manner that avoids the entrapment of water. The more the presence of water, the higher the rate of corrosion. The design also has to be made in a way to allow protective coatings to further prevent rust. Another way in which rust can be limited and prevented is by the use of inhibitors. Inhibitors are chemicals, which are injected into the iron and they prevent the formation of corrosion cells. Some examples are red lead and barium metaborate.

2.5 Industrial Applications of Hardened Steels

There are many different applications for steels; in some cases, they are required to be ductile while in some others they need to be hard. As the focus in this chapter is on hardened steels, medium- and high-carbon steels are considered in this section. Medium-carbon steels are classified by having a carbon concentration equivalent to 0.3–0.6%, and 0.6–1.0% for high-carbon steels. Due to their high wear resistance on the surface and soft cores, it is evident that they are capable of serving a broad range of applications. The product that a hardened steel is used to make is directly reflective of the specific hardening process that it has undergone. Carburizing and induction hardening are the most common case-hardening processes, each playing a vital role in the successful production of many products available today.

2.5.1 Applications of Case-Hardened Steels

The prominence of case hardening as a means for processing materials is indicated by its share of one-third of the market of heat treatment industry. One of the primary uses for case-hardening steels is in automotive transmission gearboxes. Toothed gears found in most gearboxes are usually composed of 16MnCr5 20MoCr4, two of the most frequently used case-hardened steels [25]. Case-hardened steels are predominately used for gears due to the higher maximum stress limits observed at the roots and flanks of their teeth versus those processed by induction or gas nitriding. Generally, the smallest magnitude of bending and torsional stresses is generated at the core of case-hardened gears; however, the ratio between the component's surface area and thickness must still be taken into account [25]. Case hardening is the preferred method for manufacturing steel gears over alternative approaches like carburizing, induction hardening, and gas nitriding for the following main reasons. With respect to induction hardening or gas nitriding, the hardened thickness of the gear would have to precisely correspond to the anticipated amount of wear on the component's outer surface. With this in mind, it is observed that gear surfaces hardened by gas nitriding are typically too thin, while those that are induction hardened are not hard enough to withstand surface loads present when the gear is in motion [25]. Moreover, gears that are case-hardened are less vulnerable to spalling under impact and shock-loading conditions; therefore, case-hardening is the preferred heat treatment technique for gears. Case-hardened 17Cr3, due to its low-annealed strength, is ideally used for extrusion of pistons' thin pins, while a boron alloy is typically used for their thicker walls [25]. Some additional applications for case-hardened steels in automotive manufacturing are listed below [26].

- Universal joints
- Driving pinions
- Clutch plates
- Roller bearings
- Camshafts.

2.5.2 Applications of Induction Hardened Steels

The process of induction hardening is most commonly used to prepare steel components found inside vehicular internal combustion engines. The crankshaft found in such engines was one of the first parts to be made of hardened steel. Due to impact forces and shock-loading inside the engine, the selected crankshaft build material must demonstrate appropriate tensile strength and wear resistance attributes [27]. Furthermore, the attention to such material properties in design of crankshafts is continuously rising in parallel to the increasing horsepower output of modern-day engines. Induction hardening is recognized as the most suitable heat treatment process for steel crankshafts largely because of the non-uniform hardening and resulting distortion presented by alternative methods like flame hardening and gas nitriding. In the case of 1045 Steel, one of the most common materials used for crankshafts, hardness magnitudes above 55 HRC (Rockwell Hardness Scale) can be achieved after induction hardening. Moreover, a number of additional factors suggest why this process in particular is best suited for crankshaft production. To begin, in induction hardening, only segments of the steel that need to be hardened are heated. Consequently, the remainder of the crankshaft length remains relatively soft; therefore, it eases the subsequent machining and balancing procedures that must be performed. Secondly, due to the rapid heating involved in induction hardening, the degree of distortion and scaling of the steel being used is minimized. In addition, since induction heat treatment is commonly a self-automated process, a successive tempering operation can be performed immediately after the hardening process, all within the same manufacturing cell [27]. As a result, the component cycling time is reduced; thus, it increases part production output. Overall, the outstanding level of strength along with torsional and bending fatigue resistance that are produced by induction hardening substantiates its popularity in the making of engine crankshafts.

The axle shafts found in most vehicles are also the product of induction hardening. With this approach, the axle surface is subjected to a state of compressive residual stresses, which extends its bending and torsional fatigue lifetime. In fact, increases of as much as 200% in the duration of axles' safe operation life span have been seen versus those that are conventionally heat treated [27].

Induction hardening is commonly implemented to extend the service life of rolling-mill rolls. Due to the nature of the application served by rolling-mill rolls, their operational lifetime is inevitably limited by the consistent abrasive surface wear they are subjected to. As the diameter of the rolls decreases over time due to wear, adjustments in their respective positions to meet the designed rolling reduction can only be made for so long. Moreover, the induction heat treatment process must be eventually carried out on the rolls to produce a deep hardened case enclosing their outer surfaces [27].

2.5.3 Applications of Carburized Steels

In carburizing, carbon concentration of the steel surface is selectively increased to form a material with a hard, wear-resistant outer case, in combination with a tough and ductile core [28]. Although both case and carburized steels are most commonly utilized for gears, the specific type of gear that they are best suited for does in fact differ. Case-hardened steels are preferred when manufacturing more wear-resistant gears that are capable of sustaining high impact loads, most commonly observed in automotive gearboxes. On the other hand, carburized gears are more appropriate for gear trains found in large-scale systems where minimizing mechanical noise over the system's lifetime is vital. Such is in the case of wind-turbine gearboxes, which must maintain quiet operation. It is inevitable that mechanical noise will increase over the gears' lifespan due to abrasive forces between their teeth surfaces. To combat this issue, surface hardness and abrasion resistance of the gears must be increased, which is where the properties of carburized steels can be capitalized upon. Furthermore, low-carbon alloys such as 20MnCr5 that are generally used for case-hardened products are not ideal when extended fatigue life is of highest priority. However, high-performance carburized steels like NiCrMo do in fact meet this demand [28]. Carburized steels show increased toughness and tensile strength of core, higher fatigue strength of both core and case, better performance at elevated temperatures, and low distortion upon quenching. These characteristics of carburized steels maximize their use in large and heavily loaded gears. An example of this category of steels is 18CrNiMo7-6, which is the standard grade steel for gears, shafts, bearings, bushings, and heavy-duty arbors.

2.6 Challenges in the Machining of Hardened Steels

Machining of materials harder than 45 HRC (Rockwell hardness) is usually known as hard machining [29–31]. Hard machining is principally a finishing or semi-finishing process of hard materials to achieve dimensional and form accuracy as well as the great surface finish. Although the 45 HRC is the bottom border for defining hard materials, hard machining typically involves cutting materials with hardness up to 70 HRC and usually within the range if 58–68 HRC [29]. The materials typically include but are not limited to the following categories of materials:

- Hardened alloy steels
- High-speed steels
- Tool steels
- Case-hardened steels
- Hard-chrome coated steels
- Nitrided irons.

Since the introduction of hard machining as a new concept in mid-1980s, hard turning has been among the most widely used classic hard machining operations. However, the concept of hard machining is not limited to hard turning while it can be performed in almost any other conventional machining operations such as milling, boring, and broaching. These processes are called hard milling, hard boring, and hard broaching accordingly.

Applications of hard machining are broadened by growing market demands for high production rates while maintaining the production cost as low as possible. One of the possible options for decreasing the production cost is eliminating the costly operations and replacing them with a more cost-efficient process. In such cases, the alternative process must be able to generate the required profiles with great dimensional accuracy and surface quality.

The cost-efficient machining becomes an issue when it comes to the machining of hard materials due to their extreme hardness and difficulty of cut. The common process to achieve a great surface finish in machining of hard materials is grinding which is a relatively expensive and time-consuming process. Moreover, the frequently used grinding wheels are not able to machine the complicated contours; hence, the form grinding wheels must be utilized in such cases, which impose extra cost and time to the process.

Although grinding is the common solution to finishing hard material, especially for components made of hardened steel, there are several advantages to the machining of hard materials by means of conventionally used cutting tools. Machining can be an efficient alternative for grinding if the desired dimensional accuracy and surface quality can be achieved. Replacing grinding by machining reduces the machining time, machining sequence, and also the specific cutting energy [24]. The following can be mentioned as the general advantages of hard machining in comparison to grinding operation:

- Hard machining is a popular and economic process for producing parts with high surface finish.
- It can produce surface finishes close to grinding finish.
- Hard machining is more flexible, less time-consuming, and economically feasible in comparison to the grinding.
- Although grinding is extensively used in industry for the finishing process, in many cases it can be replaced by hard turning which is a better option for internal and external finishing.

More specifically, hard machining presents the following benefits over grinding operation [29]:

- Complex geometries can be easily machined rather than being ground.
- Loading and unloading parts can be performed faster in machining.
- Several machining operations can be performed in one setup of machine tools, which may need several setup fixtures in grinding.
- Metal removal rate in machining is higher than grinding.
- Hard machining is more environmentally friendly than grinding as it can be done in the absence of coolant in many cases.
- Machine tool investment and cutting tool inventory in machining is lower than those of grinding.

Replacing grinding by machining is favorable, especially when it comes to machining of hard materials; however, the limitation and drawbacks of such replacement must be also considered. Despite its above-mentioned advantages and benefits, hard machining has its own limitations that can be listed as follows [29]:

- The price of tool per unit in hard machining might be higher than that of grinding.
- The part must have enough rigidity for hard turning. Length–diameter (L/D) ratio no more than 4/1 is preferred.
- The machine must be rigid enough to achieve the certain level of accuracy.
- The machine rigidity is an issue especially when tighter tolerances and finer surface finishes are required.
- Surface quality of part decreases by increasing the tool wear.
- White layer may be formed during machining which adversely affects the machined surface.

Despite all the above-mentioned benefits, machining hardened steel can be challenging due to the high concentration of carbon in the alloy. The main factors that would represent a challenge when machining hardened steel are high temperature generated during machining and appropriate cooling techniques. Tool wear, material selection of the tool, angle of contact and cutting edge geometry, cutting speed and force are other important factors to be considered in machining hardened steels [32]. Those factors can have positive or negative impacts on the machining process, but the negative impact is usually higher due to the brittleness of the material. When machining hardened steel, the processes must undergo the minimum amount of vibration to avoid cracks and fractures. All of the previously mentioned challenges can be analyzed in more detail, in order to predict and simulate the conditions to machine different families of hardened steel.

2.7 Hard Turning

Hard turning is known as an economic way to generate a high-quality machined surface. Typically, hardened steel parts are finish machined using grinding process, which is known by its relatively low metal removal rate. Since mid-1980s, hard turning has been one of the most commonly used classic hard machining operations [29]. It is a single-point turning of hardened steels (with hardness more than 45 HRC and usually beyond 50 HRC), and it is usually considered as a finishing process [33]. High hardness of workpiece and type of cutting tools used in this process are the major differences between hard turning and conventional turning operation. Hard turning is an alternative for grinding operation during the finish machining of hardened steels and/or other hard materials. It introduced several major benefits in comparison to grinding such as lower investment and cost of required equipment, lower amount of machining sequence and shorter setup time, comparatively higher metal removal rate, improved surface quality, and last but not least, the possibility for the elimination of cutting fluid [33–36].

2.7.1 Hard Turning as an Alternative for Grinding

Although the cutting tools for hard turning are usually costlier than a simple grinding wheel, when it comes to machining complex contours using multi-point contact form grinding wheels, hard turning shows its great advantages. Offering process flexibility and high quality, hard turning has been proven to reduce machining time; thus, it can be considered as a less costly alternative to grinding [37]. Using advanced programmable machine tools, turning tools can be programmed to cut profiles that would require time-consuming dressing routines on a grinding wheel. Making judgment about hard turning as an alternative for grinding must be based on their similarities and differences such as:

- The common high negative rake angle in hard turning produces a compressive stress field similar to that produced in grinding.
- The structural changes beneath the machined surface are likely to occur in hard turning similar to grinding.
- In both processes, the tensile residual stresses, which are induced on the outer surface of the workpiece, are attributed to the high temperature generated during machining.

Despite many similarities, there are several distinctive differences between hard turning and grinding such as:

- The contact length between tool and workpiece in hard turning is much smaller than that of grinding; as a result, the average stress in hard turning is higher than that of grinding.
- The compressive stress in hard turning occurs over the entire wear land, but in grinding, it occurs in the grain contact area.

Figure 2.3 compares hard turning and grinding as two alternatives from financial point of view.

In Fig. 2.3, the overall area of each circle shows the total cost of the corresponding process. As can be seen, replacing grinding by hard turning offers many cost advantages. As a machining operation, a single-point turning tool can simply machine complex profiles at a higher metal removal rate without need for the expensive form grinding wheels. In addition, as the majority of advanced machine tools are equipped with multiple tools, replacing grinding by hard turning allows the manufacturer to perform multiple operations using only one setup. Having one setup without requiring changeover will result in excellent dimensional and geometrical accuracy. It also eliminates the part handling time and reduces the risk of damage during handling [38].

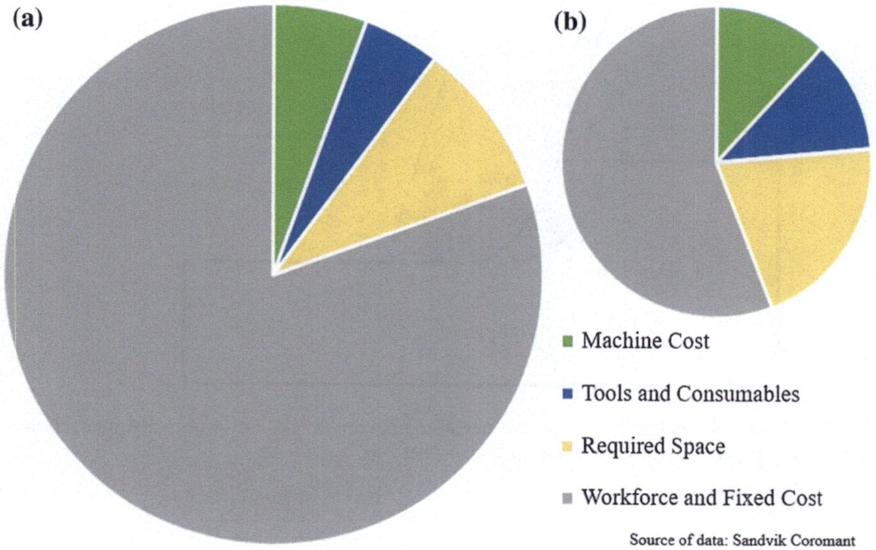

Fig. 2.3 Estimation of cost for **a** grinding versus **b** hard turning

2.7.2 Special Features of Hard Turning

As a process, hard tuning is a typical finish turning operation of hardened steel parts. However, because of the characteristics of hardened steels and recommended tool geometries, hard turning has special features that distinguish it from typical turning operations.

- Under all the cutting conditions during hard turning, saw-toothed chips are formed and represent the main feature of the process.
- Cutting tools with negative rake angle are used in hard turning to avoid chipping of the tool tip (see Fig. 2.4).
- Cutting tools in hard turning are commonly chamfered and honed.
- The ratio between the thrust and cutting forces generated in hard turning is more than 1. It is important to mention here that in a typical turning process, cutting force component is much higher than the feed and thrust component.
- The specific energy consumed in hard turning is much less than that consumed in grinding process.
- As a dry machining process, hard turning offers an alternative approach to eliminate the use of coolant, thus provides an environmentally friendly process.
- Hard turning was proven to be more economical as it shortens the machining time and production sequence. Figure 2.5 compares the conventional finishing operation of a hardened steel ring using grinding with optimized finishing of the same part using hard turning process. As it can be seen in Fig. 2.5, the optimized process not only offers time saving but also eliminates some heat treatment process and reduces environmental footprint [39].

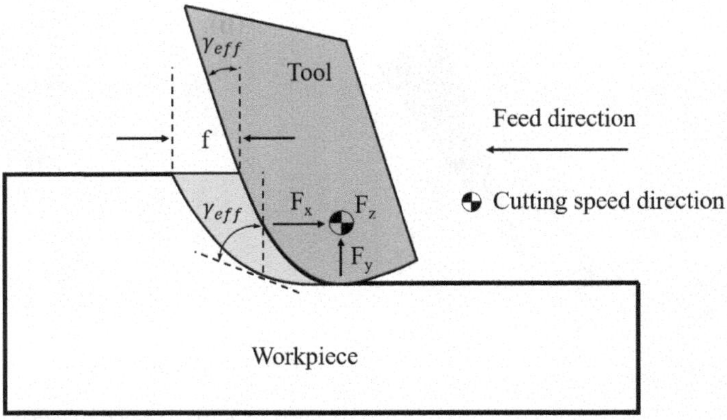

Fig. 2.4 Schematic diagram showing the negative rake angle and tool–workpiece interaction in hard turning process (with permission to reuse) [40]

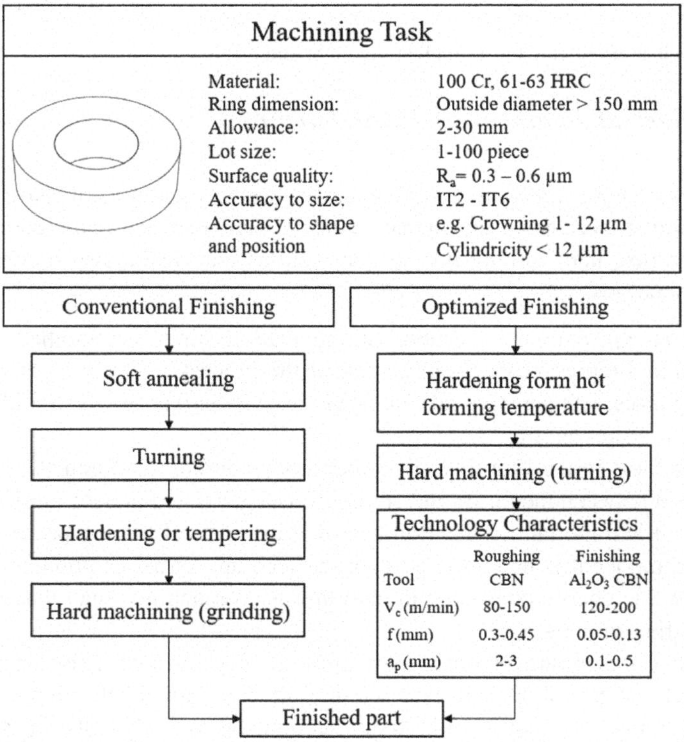

Fig. 2.5 Comparison between traditional and optimized processes for finish machining of a hardened steel ring (with permission to reuse) [39]

2.7.3 *Rigidity Imposed Limitations in Hard Turning*

Due to the nature of the process, hard turning is not applicable for the parts with low rigidity, as they cannot resist against the cutting forces during the process. Length-to-diameter (L/D) ratio is a good measure to determine whether a part is appropriate for hard turning or not. Typically L/D ratio of less than 4/1 for unsupported workpieces [29] and less than 8/1 for supported workpiece is recommended. Chatter is likely to occur for longer parts or for parts with L/D ratio of higher than above-mentioned limits.

In addition to part rigidity, machine rigidity also plays an important role during hard turning. In many cases, especially when the workpiece is rigid enough, the accuracy of hard-turned part is determined by the rigidity of machine tool. Machine rigidity is very significant, when tighter tolerances as well as finer surface are required [29]. To achieve the best cutting performance (dimensional accuracy, surface finish, longer tool life), the turning machine used in hard turning must be well designed and have a rigid structure. To satisfy the rigidity requirements, several solutions may be considered. These solutions include but are not limited to reinforced machine bases, spindles with direct-seating colleted system to position the spindle bearing as close as possible to the workpiece, and finally hydrostatic ways [29]. In addition, the vibration can be reduced using superfinished tracks, heavy-duty linear guides, and centrally located, short-pitch ball screws.

2.7.4 *Surface Quality and Integrity*

Widespread application and popularity of hard turning are adversely affected by some surface quality and integrity issues. In this section, some commonly observed surface integrity issues in hard turning are presented.

2.7.4.1 Formation of White Layer

White layer is an undesirable microstructural change near the surface region of the workpiece [41]. White layer is likely to be formed in both hard turning and grinding operations. It is an undesirable microstructural change and often called white layer as it appears as a white layer near the free surface of the workpiece [41]. Machined surfaces generated with worn tools are susceptible for white layer formation.

The white layer cannot be visually detected by naked eye, and it only appears under metallographic examination. The depth of white layer can widely vary within the range of few microns to few tens of microns [41–43]; however, generally its thickness is in the order of 1 micron. Formation of white layer is not desirable when the part is going to be used under high contact stresses or when fatigue characteristic is important. Figure 2.6 demonstrates an SEM image of a typical subsurface microstructure of a hard-turned D2 steel bar, and Fig. 2.7 shows a typical example of microstructural changes at a hard-turned surface of 52100 steel.

It has also been observed that the formation of white layer is affected by cutting conditions [41]. In general, white layer can appear by the following main reasons [42, 44]:

- Severe plastic deformation that causes rapid grain refinement, particularly when a wear land is formed along the flank face.
- Phase transformations because of rapid heating and quenching.
- Chemical reaction with tool and/or environment.

In comparison to the first two mechanisms, the last one is not likely to happen. In addition, the effects of the first two mechanisms are not easy to distinguish, as they are interdependent. The above-mentioned reasons can be described using the tool–chip cross-sectional view; see Fig. 2.8.

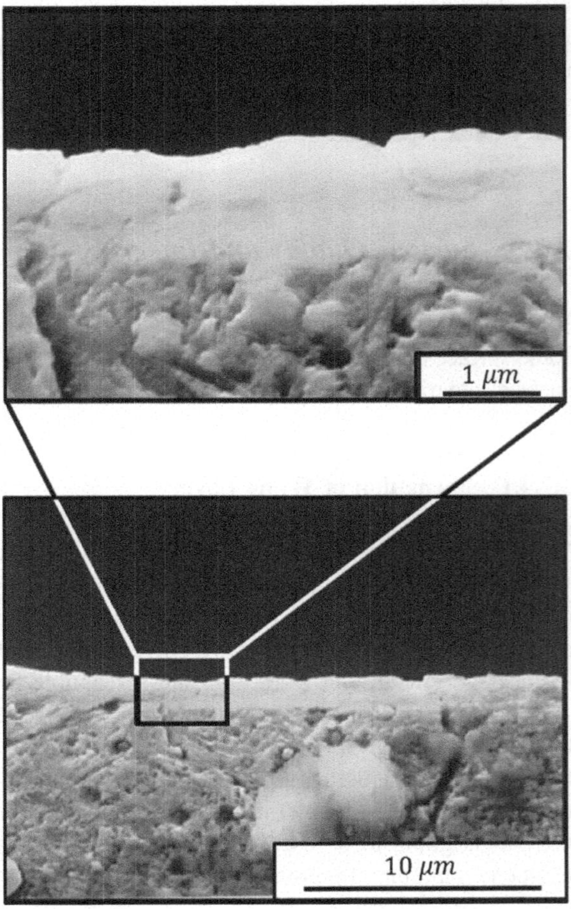

Fig. 2.6 Formation of white layer on the free surface of a hard-turned D2 steel bar (V = 270 m/min, f = 0.1 mm, d = 0.2 mm, worn tool with 1.2 mm nose radius [24]

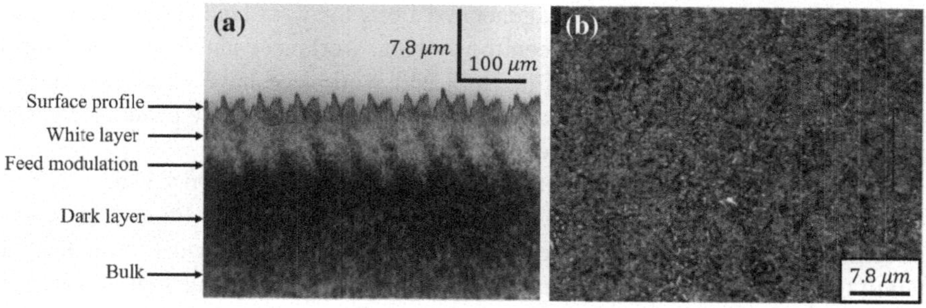

Fig. 2.7 **a** An example of microstructural change at a hard-turned surface of 52100 steel, optical micrograph, and **b** bulk microstructure (with permission to reuse) [42]

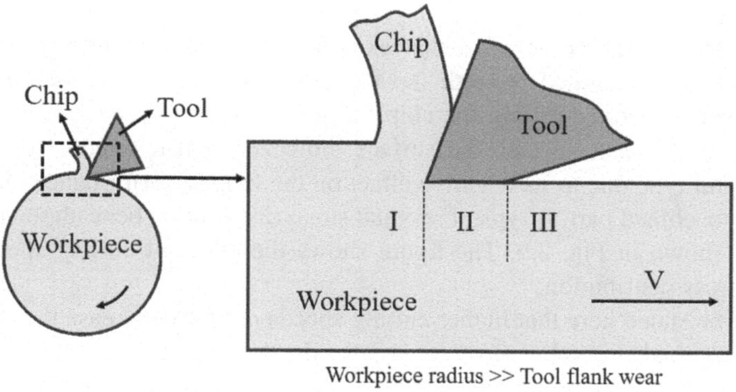

Fig. 2.8 Typical tool–chip cross section in a turning process (with permission to reuse) [42]

As can be seen in Fig. 2.8, the relative position of tool and workpiece can be presented by three distinct regions [42]:

- Region I: where the workpiece material is ahead of the cutting tool
- Region II: when the workpiece material is already passed the cutting edge and is in contact with the flank face of the cutting tool
- Region III: when the workpiece surface leaves the contact area.

In Region I, the cutting tool is pushed against the workpiece to remove a layer of material and performs the cutting action; therefore, the workpiece material in this region is severely compressed and plastically deformed. In this region, the heat generated during the cutting action is propagated from the shear zone to the tool and workpiece [42].

In the Region II, known as wear land area, the workpiece material is rubbed against the flank face of the cutting tool. This rubbing generates significant amount of heat, which is again transferred (by conduction) to the workpiece body and the cutting tool. The tool–workpiece contact area in this region is analogous to a moving heat source over a cylindrical surface. Since the workpiece is a more effective heat sink, most of the heat generated along the wear land is transferred into the workpiece. Close contact between the tool and workpiece and high temperature may initiate chemical reactions if the tool and workpiece have chemical affinity [42]. In addition, the thermal load along this zone generates tensile residual stresses within upper layer of the workpiece surface.

In the Region III, several events are likely to occur as the workpiece material leaves the contact region. These events include, but are not limited to stress unloading, cooling by means of the environment, coolant (if applied), and unheated part of the workpiece [42]. Chemical reaction is also possible at this region once the heated workpiece surface is exposed to the environment.

2.7.4.2 Residual Stresses

Residual stresses are the stresses that exist in a part after removing all external loads. They are generated beneath the machined surface due to mechanical and thermal loads generated during the chip formation process. Both tensile and compressive stresses exist beneath the surface; however, tensile residual stress is the most harmful type due to its negative effect on the fatigue performance and service life of the machined part. A typical residual stress distribution beneath the machined surface is shown in Fig. 2.9. The figure shows the effect of cutting speed on the residual stress distribution.

It must be stated here that higher cutting speeds not only increase the magnitude of tensile residual stress, but also increase the depth of residual stress layer beneath the surface. As mentioned before, cutting with worn tools also promotes the formation of tensile residual stress. In finish machining process, the end of tool life is

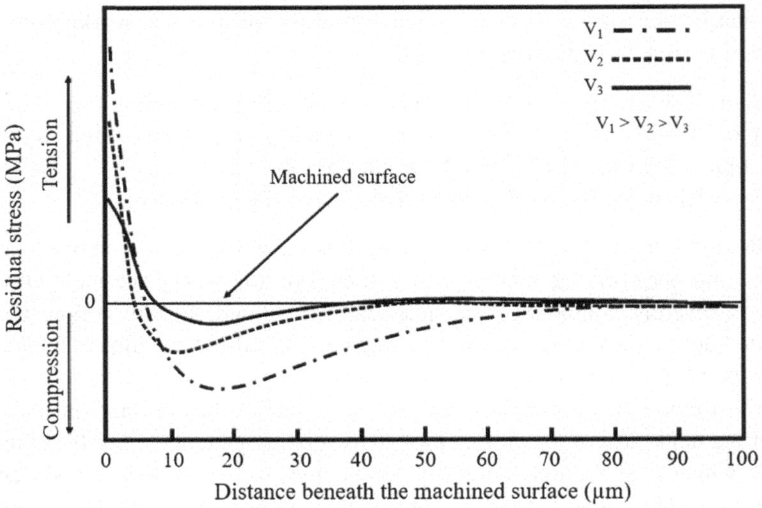

Fig. 2.9 Typical distribution of residual stress beneath the machined surface in hard turning

governed by the maximum wear land value. The maximum wear land is not only selected based on the ability of tool to continue cutting but is based on a maximum allowable tensile stresses generated on the surface.

2.7.4.3 Material Side Flow

Examining a machined surface under the microscope shows material side flow as a burr formed along the feed marks rides. Side flow is formed because of squeezing the material under high temperature and pressure between the tool tip and the machined surface. Material side flow is not only formed during hard turning, but also is formed when the generated pressure and temperature are enough to cause viscous fluid like material behavior. Figure 2.10 shows a typical material side flow observed on hard-turned surfaces.

The mechanism of forming material side flow is shown in Fig. 2.11. When hard turning is performed using a tool with nose radius, the thickness of the chip varies along the curved part of the cutting edge. The variation is such that there are some areas where the chip thickness is below the minimum chip thickness needed to form a chip. In these areas (right hand side in Fig. 2.11), the material will be squeezed and spreads to the sides along the feed marks instead of being cut.

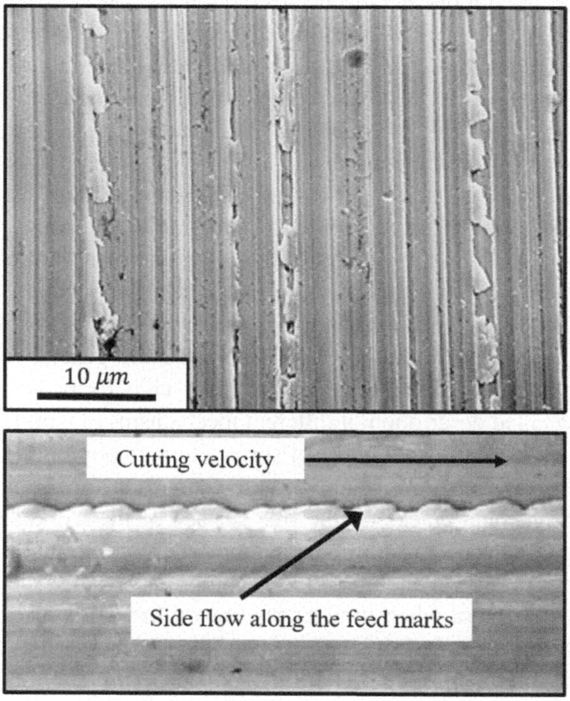

Fig. 2.10 SEM images of material side flow when turning case-hardened steel AISI 4615 (60 HRC) (with permission to reuse) [40]

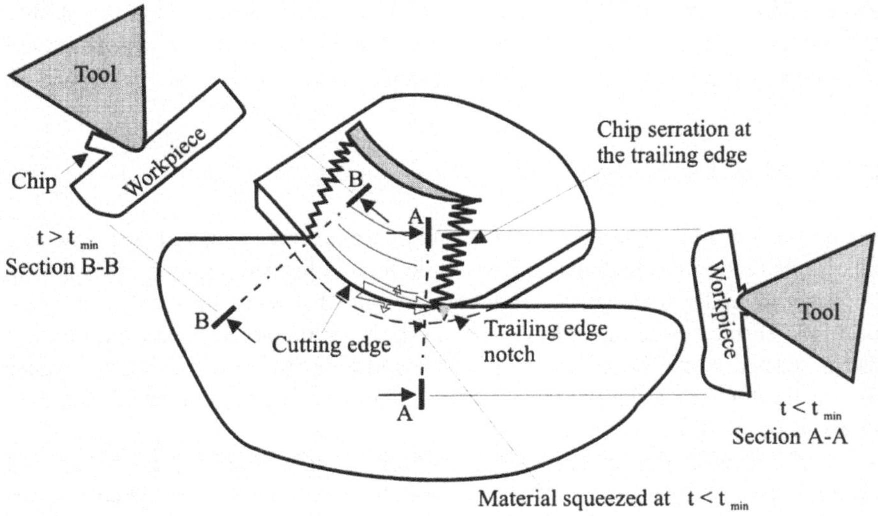

Fig. 2.11 Mechanism of side flow during hard turning (with permission to reuse) [40]

2.8 Mechanics of Chip Formation During Hard Turning

In order to study machining of hardened steels, their cutting characteristics, and required tools, the mechanics of chip formation must be investigated. Mechanics of chip formation is greatly affected by the surface topography and properties of material to be cut. In the case of ductile materials where the part is relatively soft, the chip is formed by severe plastic deformation of workpiece material in the shear zone [23, 45]. In contrast with ductile materials, the mechanics of chip formation when machining hard and relatively brittle materials is highly affected by the surface topography of workpiece [24] and hardened steels are no exception. In such cases, the mechanics of chip formation is dominated by crack initiation at the free surface or in the vicinity of tool tip rather than plastic deformation. However, some level of plastification is still observed when studying the chip cross-sectional area [39]. The cracks are initiated where no hydrostatic pressure exists such as workpiece surface [23]. Figure 2.12 shows the general model of continuous chip formation for ductile and saw-toothed chip formation for brittle material. A more detailed description of different mechanisms of chip formation can be found in [45].

The significant effect of surface topography on the mechanics of chip formation in hard turning is mainly due to the presence of irregularities such as microscopic ridges and cracks on the free surface of the workpiece. Figure 2.13 demonstrates the mechanism of chip formation when turning hardened steels.

As shown in Fig. 2.13, a compressive stress field is generated ahead of the cutting tool when it advances toward the workpiece. The compressive stress filed along with the brittleness of workpiece material initiate a crack on the free surface of the workpiece [34, 46]. The crack initiation is owing to irregularities and

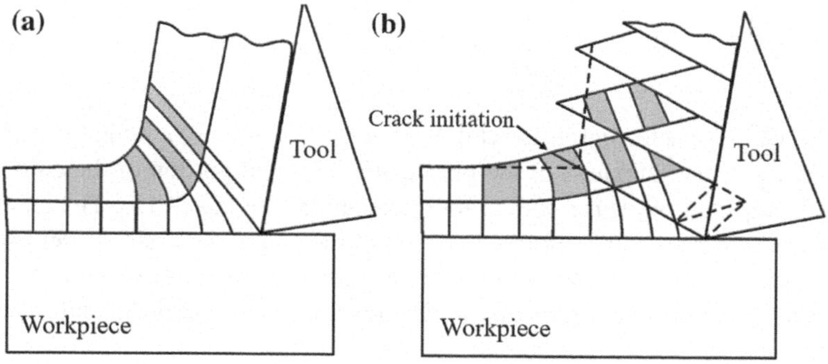

Fig. 2.12 Models of chip formation. **a** Continuous chip and **b** saw-toothed chip (with permission to reuse) [23]

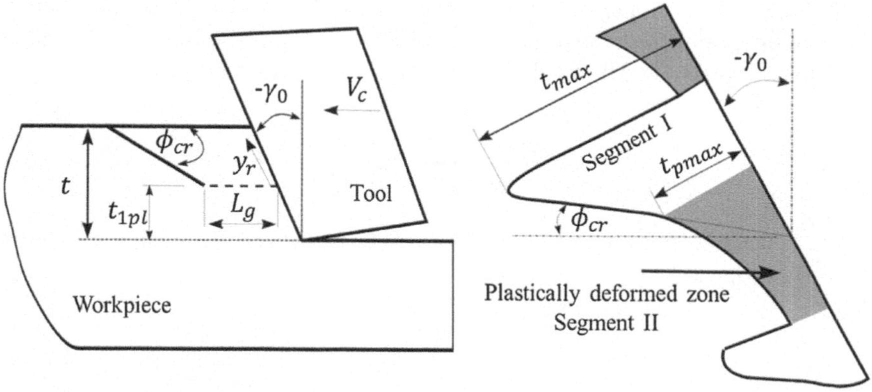

Fig. 2.13 Mechanism of chip formation when turning hardened steels [24]

microridges exist on the free surface of the workpiece. As the cutting tool continues to move forward, the crack propagates toward the tool tip. However, the presence of severe plastic deformation in the close proximity of cutting edge, due to the combined effects of high compressive strain field and elevated temperature, prevents the crack from reaching the edge. Upon further advancement of the cutting tool, the fractured material between the rake face and the crack (Segment I in Fig. 2.13), slides upward utilizing the crack as a sliding plane. Subsequently, plastically deformed region (Segment II in Fig. 2.13) is extruded through the opening between the base of the crack and rake face of the tool. The mechanics of chip formation when machining hardened steels is represented in terms of the formation of segments I and II.

The extrusion of Segment II (plastically deformed portion of workpiece material) through the gap between the rake face and the base of the crack can be articulated as follows [46]:

$$\frac{\mathrm{d}t_{\mathrm{pl}}}{\mathrm{d}y_r} = -\frac{\cos(-\gamma_0)}{t_{1\mathrm{pl}}} t_{\mathrm{pl}} \tag{2.1}$$

In Eq. (2.1), t_{pl} is the deformed chip thickness at any point y_r along the tool–chip interface, $t_{1\mathrm{pl}}$ is the thickness of the plastically deformed portion of the chip where the crack stops, and γ_0 is the cutting tool rake angle (see Fig. 2.13). The maximum thickness of the plastically deformed portion of the chip (where plastic deformation starts) $t_{p\max}$ occurs when the chip begins to move along the rake face at $y_r = 0$. Parameter y_r can also be interpreted as the traveling distance of the chip along the rake face.

$$t_{p\max} = \frac{t_{1\mathrm{pl}} \cos(\gamma_0 - \phi_{\mathrm{cr}})}{\sin \phi_{\mathrm{cr}}} \tag{2.2}$$

In Eq. (2.2), ϕ_r represents the inclination of the crack against the cutting direction or simply crack propagation angle. Solving Eq. (2.1), by using the initial conditions, reveals the deformed chip thickness t_{pl} as presented below:

$$t_{\mathrm{pl}} = \frac{t_{1\mathrm{pl}} \cos(\gamma_0 - \phi_{\mathrm{cr}})}{\sin \phi_{\mathrm{cr}}} \mathrm{e}^{\left(\frac{-\cos(-\gamma_0)}{t_{1\mathrm{pl}}} \times y_r\right)} \tag{2.3}$$

The initial gap length L_g (see Fig. 2.13) can be calculated using initial condition at $t = 0$ as follows:

$$L_g = \frac{t_{p\max}}{\cos(-\gamma_0)} = \frac{\frac{t_{1\mathrm{pl}} \cos(\gamma_0 - \phi_{\mathrm{cr}})}{\sin \phi_{\mathrm{cr}}}}{\cos(-\gamma_0)} = t_{1\mathrm{pl}} \left[\frac{\cos(\gamma_0 - \phi_{\mathrm{cr}})}{\sin \phi_{\mathrm{cr}} \cos(-\gamma_0)}\right] \tag{2.4}$$

Using Eqs. (2.3) and (2.4), the movement of chip along the rake face can be obtained as [24]:

$$y_r = \frac{t_{1\mathrm{pl}}}{\cos(-\gamma_0)} \ln\left(\frac{L}{L_g}\right) \tag{2.5}$$

Finally, the velocity of the plastically deformed chip when traveling over the rake face is expressed as follows:

$$V_P = \frac{\mathrm{d}y_r}{\mathrm{d}t} = \frac{t_{1\mathrm{pl}}}{\left[L_g - V_c t\right] \cos(-\gamma_0)} V_c \tag{2.6}$$

At $t = 0$, when the chip is about to slide over the rake face, the velocities of the two distinct segments of the chip (fractured Segment I and plastically deformed Segment II in Fig. 2.13) are equal; thus, Eq. (2.6) can be rewritten in terms of:

$$V_P = V_R = \frac{t_{1\mathrm{pl}}}{L_g \cos(-\gamma_0)} V_c \tag{2.7}$$

In Eq. (2.7), the velocity of the fractured Segment I is represented by V_R while the velocity of the plastically deformed Segment II is expressed by V_P [24]. According to the theory presented above, during the formation of chip when machining hardened steels, the maximum thickness of the deformed chip $t_{\max}$ linearly decreases until it reaches the maximum thickness of the plastically deformed chip $t_{p\max}$. In addition, as per Eq. (2.3), the thickness of plastically deformed chip t_{pl} exponentially decreases within the portion of chip that is plastically deformed (Segment II). Ultimately, the thickness of deformed chip t_{pl} approaches zero and yields a discontinuous chip [24]; however, before reaching a zero thickness, another crack is initiated and this process is cyclically continued resulting in the formation of connected saw-toothed chips.

Another important factor that affects the mechanics of chip formation when machining hardened steel is the tool rake angle. Using the negative rake angle for the cutting tool in machining of hardened steels is very common. When the tool has negative rake angle, the chip cannot flow easily along the rake face, and it is hardly pushed against the tool, which consequently generates high compressive stresses both on the tool and workpiece [39]. As previously mentioned, these compressive stresses result in the formation of crack instead of leading to material flow. Once the level of shear strain caused by stresses reaches a certain limit that can no longer be tolerated by the workpiece surface, the crack plays the role of sliding surfaces at which the stored energy is released and eventually the chip is formed. The process will be repeated after the current segment of chip has slid away. In this case, the cutting pressure generates new cracks and consequently new chips [24, 39, 45]. Figure 2.14 demonstrates the effects of the negative rake angle on the shape and geometry of the saw-toothed chip when machining hardened bearing steel.

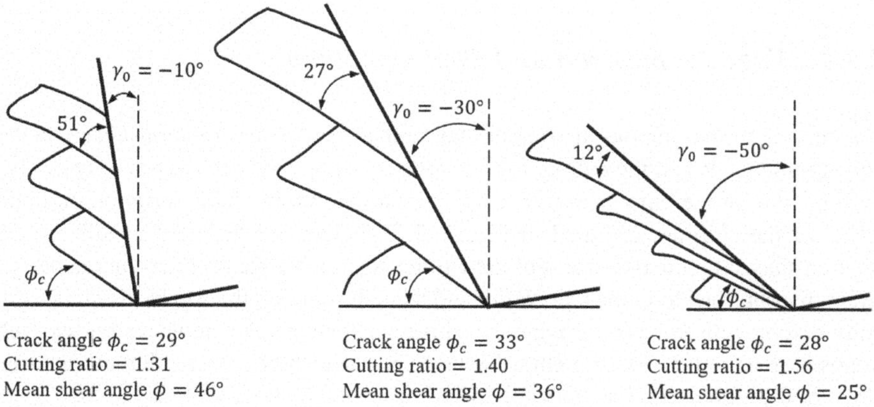

Fig. 2.14 Cross sections of chips produced in the machining of hardened bearing steel with different negative rake angles (with permission to reuse) [23]

2.9 Influential Factors on Chip Formation During Hard Turning

Several factors influence the mechanics of chip formation during hard turning among them; the following factors are the most important ones.

2.9.1 *Nose Radius*

The nose radius of cutting tool significantly affects the chip formation and changes the morphology of produced chip. The SEM images of chips collected after orthogonal turning of AISI 1550 hardened steels (58–60 HRC) show no evident edge serration on the transverse section of the chip. However, SEM images of the chips formed when tool with nose radius or round insert was implemented show a different pattern. In the case of nose radius or round insert, significant serration can be observed along the trailing edge of the chips. Figure 2.15 shows SEM images of the free surface of the chips when different tool geometries are utilized.

Serration of the trailing edge of the chips when machining hardened steels with nose radiused or round tools can be attributed to the variation of chip thickness as presented in Fig. 2.16.

As can be seen in Fig. 2.16, in the absence of nose radius, the chip thickness remains constant along the cutting edge (*A-B*), whereas in the presence of nose radius or round tool, thickness continually changes along the edge (*A-B*). Chip has its lowest thickness (almost zero) at point A that causes stress concentration. In addition, due to the radius of the cutting edge, chip velocity varies across the thickness of the chip, which leads to non-uniform displacement along the chip width and makes the transverse surface of the chip serrated.

2.9.2 *Edge Preparation and Tool Condition*

Chamfering the cutting tool also affects the mechanics of chip formation and resultant chip geometry. When the cutting tool is virtually sharp, serration exists only on the trailing edge of the chip; however, using a chamfered tool yields serration on both edges of the chip as presented in Fig. 2.17. This can be attributed to differential loading caused by the roundness of the cutting edge, which leads to inhomogeneous strain distribution along the chip width and tears the chip on the weakest side. When cutting tool with chamfered edge is utilized, effective rake angle increases and imposes severe stress on the chip. Being exposed to high stress, ultimate tearing stress of the chip exceeds at both edge yields serration on both sides. It has also been observed that when machining is performed using a worn tool, the saw-toothed chip will be smaller similar to those of the continuous chips [24].

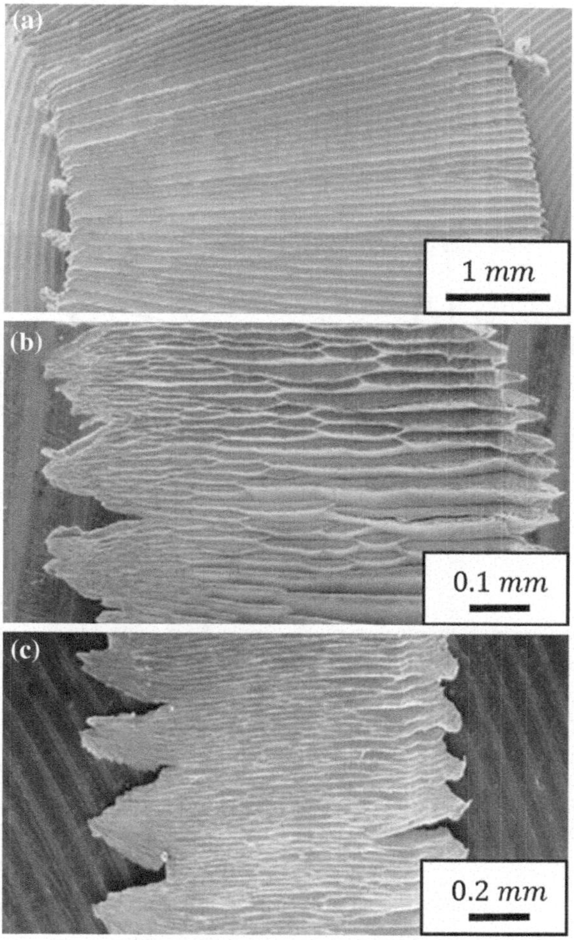

Fig. 2.15 SEM images of the free surface of the chips produced using tools with different geometries. **a** Orthogonal cutting with rake angle $\gamma_0 = -6°$, **b** oblique cutting with rake angle $\gamma_0 = -6°$ and nose radius $r = 1.2$ mm, **c** orthogonal cutting with rake angle $\gamma_0 = -6°$ using a round insert [24]

2.9.3 Feed

It has been shown previously in this chapter that when machining hardened steels, a crack is formed on the free surface of the workpiece ahead of the cutting tool and it propagates toward the cutting edge. When feed is relatively small, plastically deformed region covers a large area across the chip; therefore, it terminates further propagation of the crack. As a result, cross section of the chip is analogous to continuous chips. Figures 2.18, 2.19, and 2.20 show the effect of feed on chip morphology and chip segmentation distance when machining hardened steels.

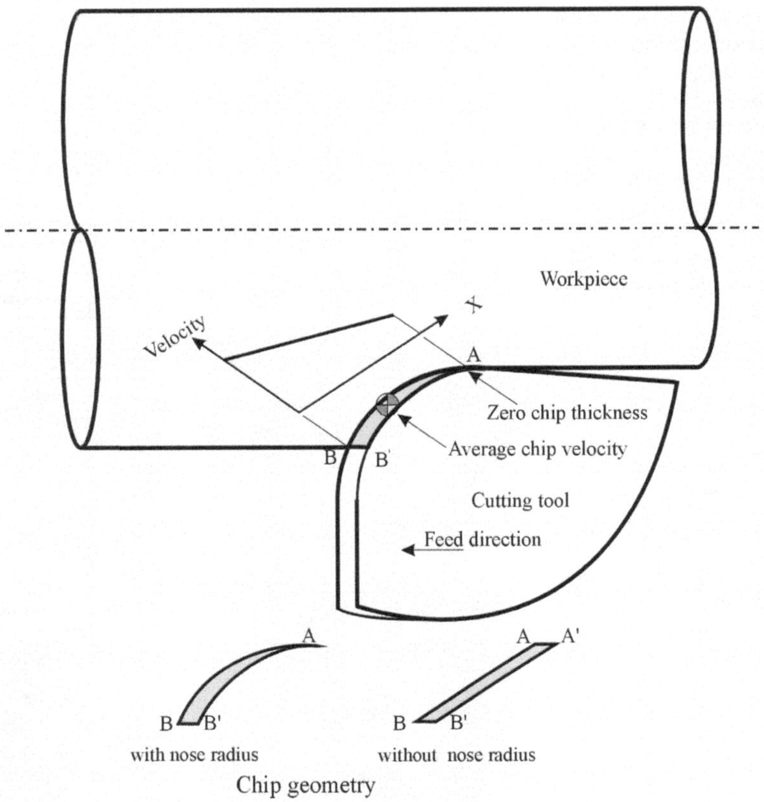

Fig. 2.16 Effect of nose radius on the morphology of resultant chip [24]

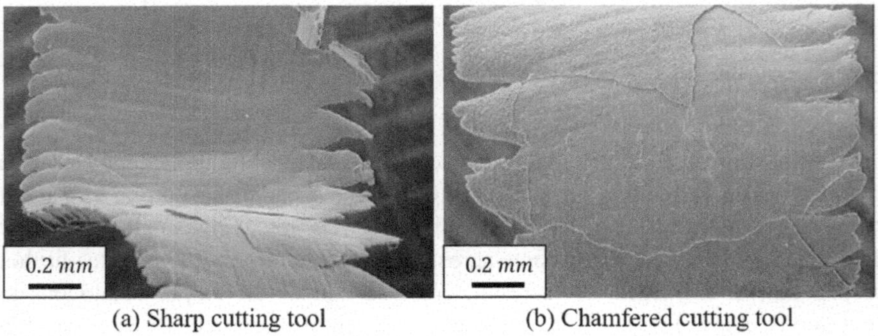

Fig. 2.17 Effect of edge preparation on the morphology of resultant chip when machining AISI 1550 hardened steels (58–60 HRC) [24]

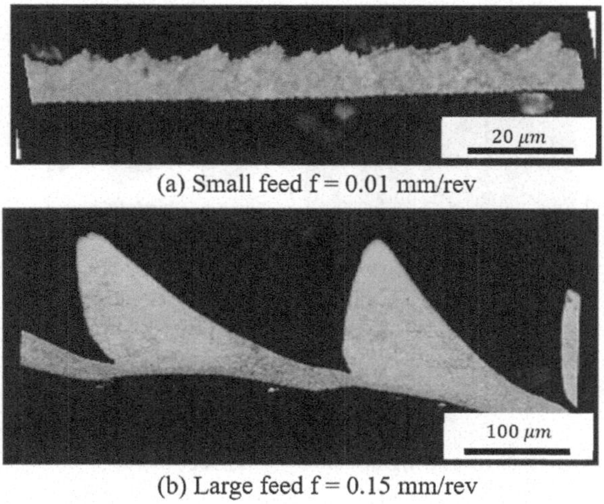

Fig. 2.18 Effect of feed on the morphology of resultant chip when machining AISI 1550 hardened steels (58–60 HRC) at v = 90 m/min [24]

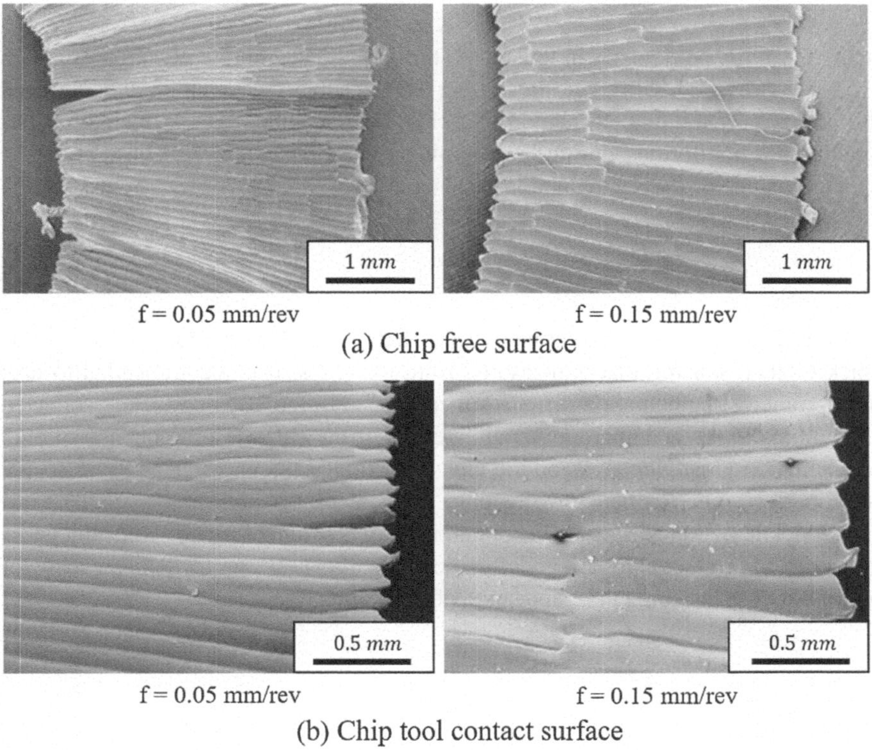

Fig. 2.19 Effect of feed on chip segmentation distance when machining AISI 1550 hardened steels (58–60 HRC) at v = 120 m/min [24]

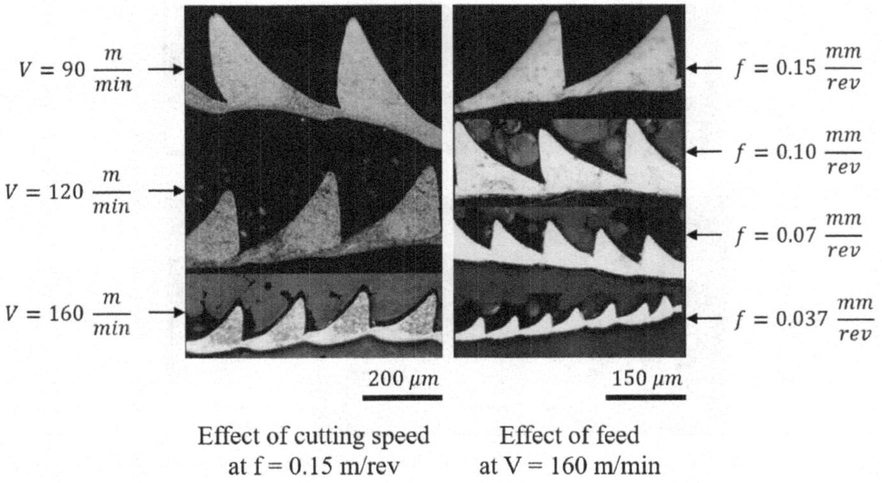

Fig. 2.20 Effect of feed and speed on chip segmentation distance [24]

2.10 Dynamics of Chip Formation

In hard turning, particularly in finish hard turning, the magnitude of cutting forces is not necessarily high; however, the cutting edge is subjected to a high dynamic load due to the periodic chip formation and chip segmentation. The segmentation frequency of chips formed during hard turning can be calculated by considering the average values of the segmentation distance by:

$$f_c = \frac{V}{\lambda_c} \tag{2.8}$$

In Eq. (2.8), V is the cutting speed and λ_c is the average value of the crack-toothed spacing that is measured from the chips. It is recommended to take the average of segmentation distances measured at different pieces of the collected chips to avoid the bedding and handling effect of the chips. The frequency of chip segmentation greatly depends on the undeformed chip thickness and cutting speed; see Fig. 2.21. The frequency of the chip segmentation is high at low undeformed chip thickness and decreases exponentially when undeformed chip thickness increases. At high values of undeformed chip thickness, or feed in the case of orthogonal cutting, the curve appears to flatten out [34].

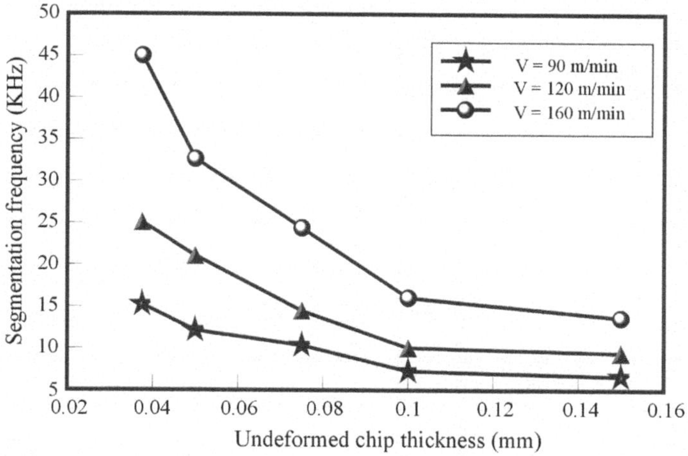

Fig. 2.21 Effect of cutting speed and undeformed chip thickness on the segmentation frequency in hard tuning [24]

2.11 Cutting Forces During Hard Turning

Despite their high hardness, the magnitude of cutting forces when machining hard materials may not be essentially high because of the following reasons [23]:

- In machining hard materials, chip is mainly formed by crack formation rather than plastic deformation.
- Friction force in machining hard material is lower than that of other materials due to the small contact area between the tool and chip.

It has been reported by several research studies [35, 47, 48] that the cutting force may be decreased by increasing the hardness. This phenomenon can be attributed to the temperature at the chip–tool interface. In the case of steel, increasing the hardness rises the temperature generated during chip formation process. This in turn contributes to the thermal softening of the workpiece material and eventually reduces the generated forces. The cutting force suddenly increases when the hardness exceeds 50 HRC which corresponds to the onset of saw-toothed chips formation [35]. In this case, the process is affected by two contradicting factors. On the one hand, the yield strength of workpiece material is increased by increasing the hardness, which requires higher forces to introduce deformation and form chips. On the other hand, the workpiece material is softened, due to the extreme heat generated during machining, and consequently yield strength is decreased. In the case of steels, if the workpiece is harder than a certain limit, as mentioned earlier, the mechanics of chip formation is dominated by crack initiation rather than plastic deformation. As a result, deformation energy, which in turn is converted to heat, is reduced and material softening does not happen. Figure 2.22 illustrates the relationship between hardness and cutting forces for steels.

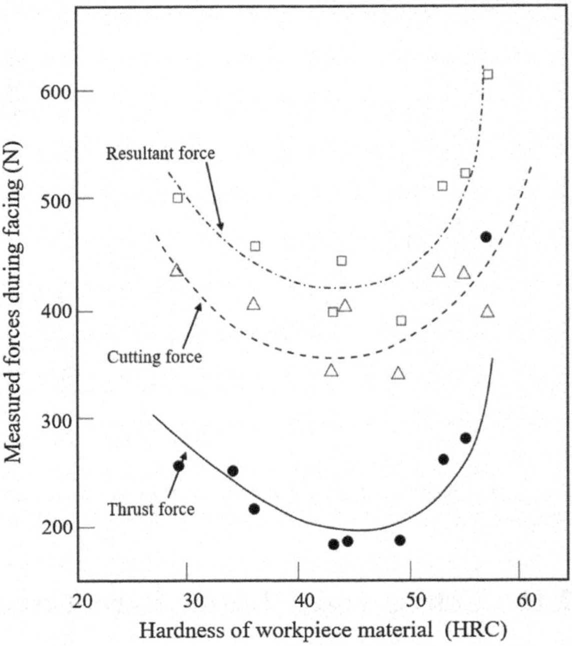

Fig. 2.22 Relationship between hardness and cutting forces for steels (with permission to reuse) [47]

As can be seen in Fig. 2.22, for soft steels, the cutting force is comparatively high and decreases by increasing the hardness. However, this trend changes once the hardness exceeds 50 HRC. Although the cutting forces are decreased to some extent by increasing the hardness up to a certain limit, the cutting tools are rapidly worn away due to the hardness of the workpiece material. The worn-out tool leads to increased cutting force specially the thrust force [24].

2.12 Appropriate Tool Materials for Hard Turning

An appropriate cutting tool to be used in hard turning must possess the following characteristics:

- Chemical and physical stability in high cutting temperature
- Hot hardness
- Wear resistance
- High toughness.

Figure 2.23 compares the hot hardness and wear resistance of several widely used cutting tool materials versus their toughness and transverse rupture strength.

As can be seen in Fig. 2.23, diamond tools are the hardest ever-existing cutting tools and they have moderate toughness and transverse rupture strength. However, despite their extreme hardness, natural and synthetic diamond can be only used for

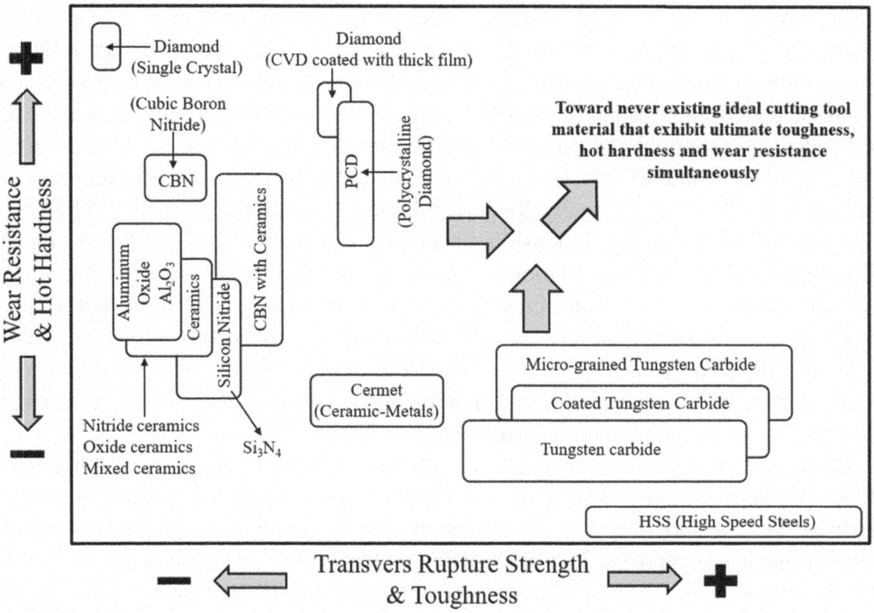

Fig. 2.23 Hot hardness and wear resistance versus toughness and transverse rupture strength for different tool materials

some few applications in hard turning such as precision turning of non-ferrous metals [29]. The diamond tools could not be used for the machining of steel, as the diamond reacts with the carbon content of steel and gets "graphitized."

Among the remaining cutting tool materials, the following are the candidates for hard turning.

- Cubic boron nitride (CBN)
- Polycrystalline cubic boron nitride (PCBN)
- Mixed (Al_2O_3-TiC) ceramics
- Cermets.

It must be mentioned that despite the fact that all of tool materials are applicable choices for hard machining in general and hard turning in particular, the appropriate cutting tool is normally selected based on the workpiece material to be cut, preferred production rates, and production cost.

2.12.1 CBN and PCBN Tools

Cubic boron nitride (CBN) and polycrystalline cubic boron nitride (PCBN) are the primary candidates for hard turning, especially when dimensional accuracy and surface quality are of great interest. These families of cutting tools are very expensive compared to other tool materials, e.g., carbide; therefore, their broad application in machining industry is always hindered. However, their great characteristics such as high hardness especially at elevated temperatures, great wear resistance, and thermal stability have made them a viable choice for hard turning [49, 50]. CBN and PCBN tools demonstrate good performance in interrupted cutting and offer comparatively longer tool life for continuous turning. They are also capable of delivering dimensional accuracy and surface quality close to those of grinding. When it comes to tool performance, tool life is one of the most widely used performance measures for cutting tools. In hard turning, tool wear is typically progressed by a combination of several wear mechanisms such as abrasion, adhesion, and diffusion [33, 51, 52]. Among the above-mentioned wear mechanisms, some of them may be more effective than the others based on the cutting parameters even for a similar combination of tool and workpiece [33].

CBN and PCBN tools are generally divided into two categories: high CBN/PCBN content (around 90%) and low CBN/PCBN content with ceramic phase, such as TiN, added [50, 53]. Members of this family with higher CBN/PCBN content are harder and yet exhibit higher toughness than those with lower CBN/PCBN content. On the other side, tools with lower CBN/PCBN content with added ceramic phase demonstrate higher chemical stability that makes them more resistant to higher temperatures and consequent diffusion wear [50, 53]. As a result, tools with higher CBN/PCBN content are generally recommended for interrupted turning of hardened steels where higher toughness is an asset. Tools of lower CBN/PCBN content are mainly suggested for continuous finish turning of hardened steels where higher chemical stability is desired. This is due to the fact that in continuous turning, tool remains in contact with the workpiece for a longer period of them and thus it will be exposed to higher temperature that promotes diffusion wear [50, 53].

Previously published research works [33, 49, 51, 52] show that during hard turning, flank wear is the main pattern of wear for CBN tools. However, crater wear is also important when it comes to effective rake angle. The rake angle is normally selected negative to give the tool enough strength to withstand against the hard material during cutting. Progressing the crater wear adversely affects the rake angle and makes it more positive which weakens the tool and is not desirable [54]. It has been shown that among the wear mechanisms, adhesion has the highest contribution (more than 80%) and abrasion and diffusion occupy the next positions [33, 55]. Figures 2.24 and 2.25 show the cutting edge of a CBN tool when cutting AISI 4340 steel with hardness of 56 HRC. These two figures show the modes of tool wear under continuous, semi-interrupted, and interrupted cutting for two grades of CBN tools: CBN 7050 with high CBN content and CBN 7020 with low CBN content and ceramic phase added. The tool represented in Fig. 2.24 has a chamfered cutting edge while the one represented in Fig. 2.25 has a round cutting edge.

As can be seen in Figs. 2.24 and 2.25, both flank wear and crater wear can be observed when CBN tool is utilized in hard turning; however, these two modes of wear are caused by different mechanisms in CBN 7050 and CBN 7020. Smooth appearance of the flank wear land for both chamfered and rounded tools made of

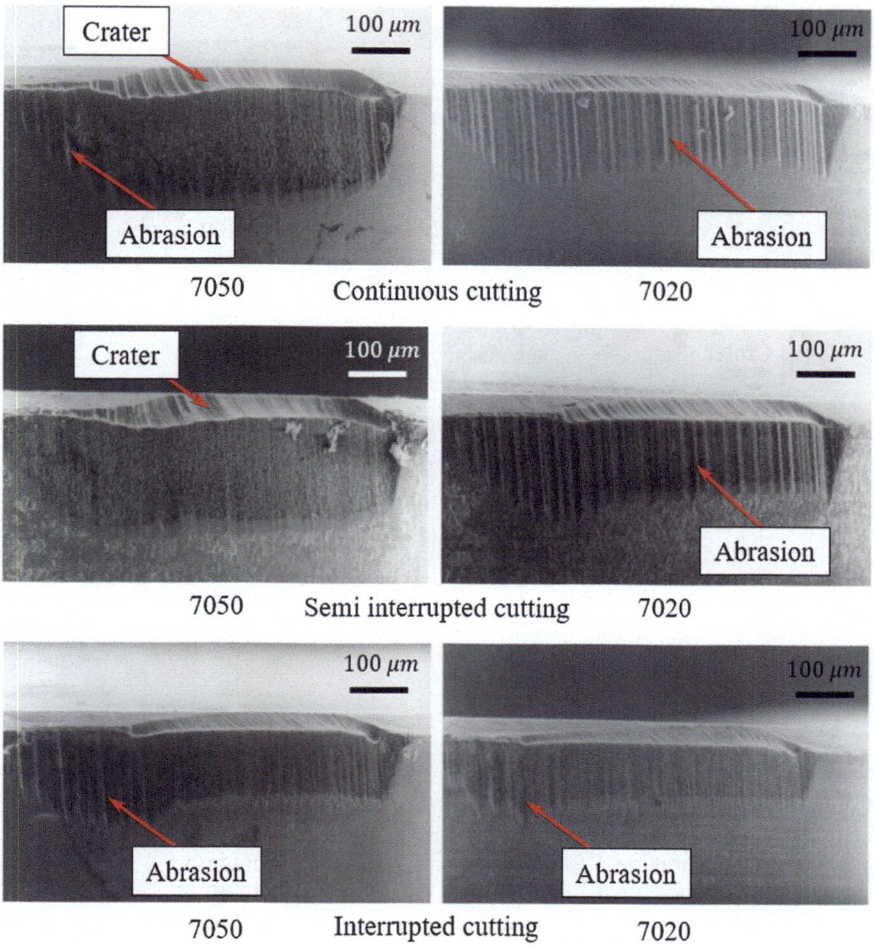

Fig. 2.24 Flank face of the CBN chamfered tool (with permission to reuse) [53]

CBN 7050 indicates that the flank wear was dominantly caused by diffusion mechanism. Same mechanism is also responsible for the crater wear in CBN 7050 tools [53]. This can be attributed to the lack of chemical stability at high temperatures due to the absence of ceramic phase. The mechanism of flank wear and crater wear for CBN 7020 is mainly abrasion rather than diffusion. This is evident by the groove marks on both rake face and flank face of CBN 7020 tools. These tools are made of lower CBN content to which ceramic phase is added. As mentioned earlier, ceramic phase reduces the toughness of CBN tool but increases their chemical stability at higher temperatures. As a result, CBN tools with lower CBN content but enhanced with ceramic phase are very resistant to diffusion wear.

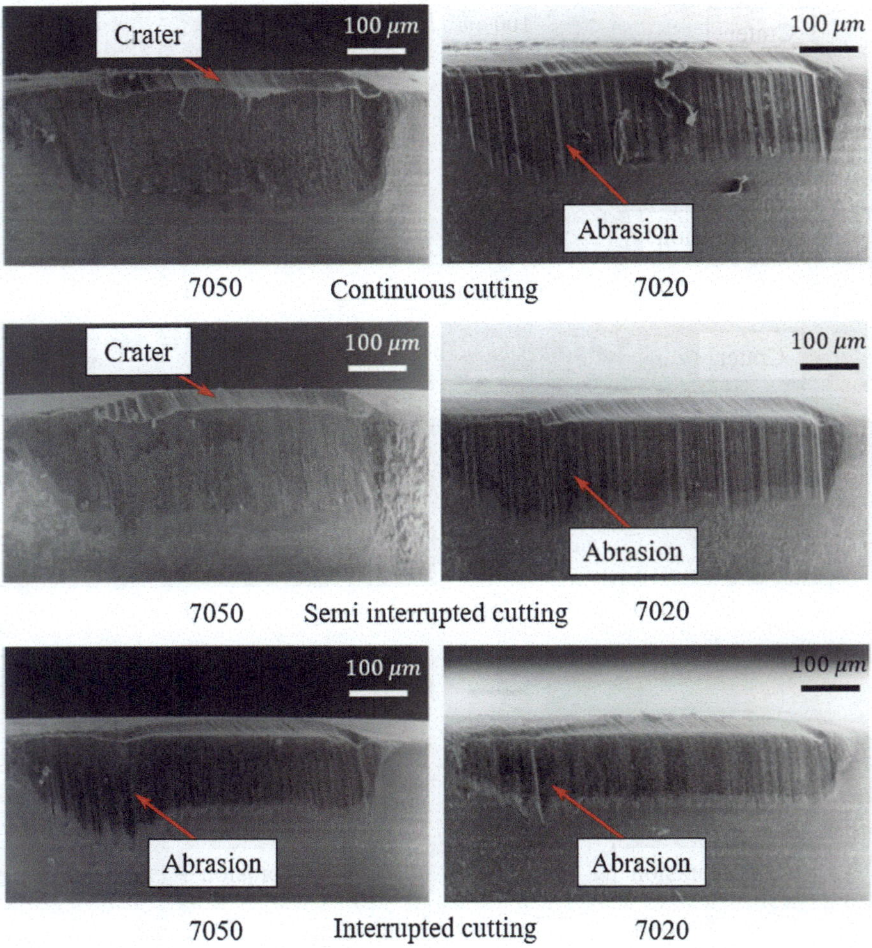

Fig. 2.25 Flank face of the CBN rounded tool (with permission to reuse) [53]

Overall, it is very clear that even in interrupted machining, the CBN cutting tool is neither chipped nor broken. This proves that CBN tools possess enough toughness to survive in hard turning even in the worst-case scenario where interrupted hard turning is applied [53].

2.12.2 Ceramic Tools

Ceramic tools are extensively used for the machining of different hard materials. They possess some favorable characteristics such as hot hardness, high melting point, and good wear resistance [56]. The cost of ceramic tools is closer to that of carbide tools that is comparatively cheaper than CBN and PCBN tools. The above-mentioned features together make ceramic tools a favorable and also economically feasible option for machining of hardened steels. Among ceramic tools, alumina-based one can be used as an alternative for CBN and PCBN tools in machining of hardened steels; however, pure alumina tools greatly suffer from lack of fracture toughness and low thermal shock resistance. As a result, they are usually used in dry cutting operations in the absence of any type of coolant. High rigidity of machining setup also plays an important role in the performance of ceramic tools and reduces the chance of fracture tool failure. The thermal shock resistance and fracture toughness of alumina-based ceramic tools can be improved by adding ZrO_2, TiC, TiN, or SiC whiskers [50].

2.12.3 *Cermet (Solid Titanium Carbide) Tools*

Cermet tools perform well for continuous turning operations. They also show good performance especially on case-hardened workpieces. Despite their tendency to wear evenly without breakage, cermet tools wear out faster than CBN tools.

According to Sandvik Coromant, the appropriate cutting tool for hard turning must be selected based on the application and hardness of the workpiece. Figure 2.26 shows the appropriate procedure for tool selection in hard turning. As can be seen in Fig. 2.26, carbide tools are not recommended if hardness of workpiece exceeds 50 HRC. Ceramic tools can be used when the hardness of workpiece is within the range of 50–60 HRC if the surface finish is not of great importance (moderate demand). Overall, CBN and PCBN tools are the ultimate choice of cutting tool for hard turning; however, they are not recommended for machining of steels with hardness level of lower than 48 HRC.

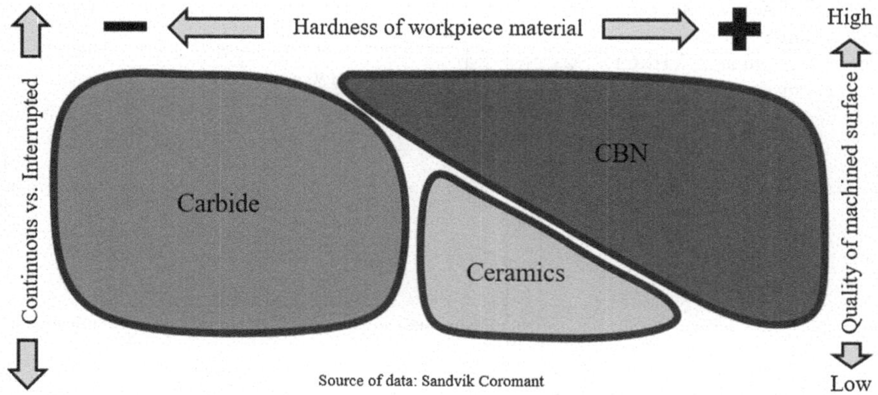

Fig. 2.26 Appropriate tool materials for hard turning

2.13 Surface Finish in Hard Turning

The main motivation for introducing hard turning to the industry was to replace grinding as an expensive finishing operation with a more viable and flexible machining operation. Hard turning must be capable of delivering equivalent surface finish to be considered as a practical alternative for grinding. It has been shown [57, 58] that surfaces finish in hard turning is dominantly affected by feed rate and geometric features of the cutting tool, mainly nose radius. Average surface roughness increases by increasing the edge radius. This can be attributed to the increase of the ploughing component of deformation in comparison to its shearing component. This effect is decreased by increasing the hardness of work material [58]. Figures 2.27 and 2.28 show the effects of feed rate and edge geometry on the surface roughness of the workpiece material.

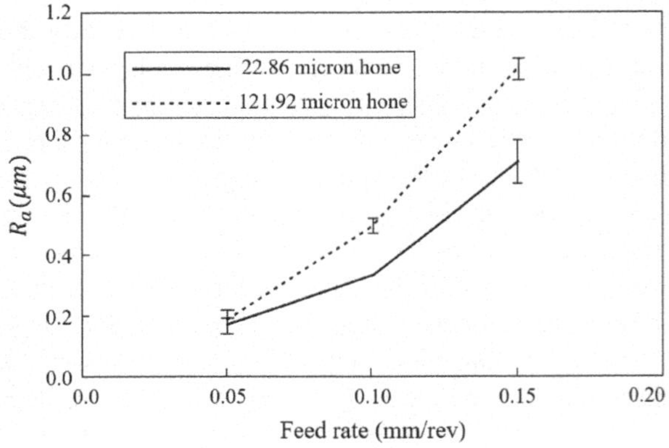

Fig. 2.27 Effect of feed and edge geometry on R_a (workpiece hardness 47 HRC) (with permission to reuse) [58]

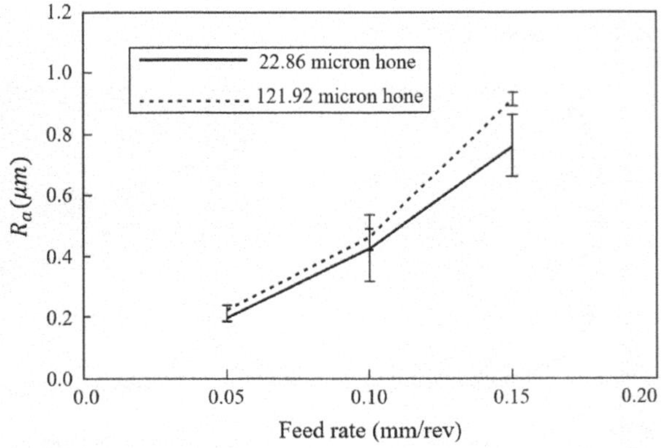

Fig. 2.28 Effect of feed and edge geometry on R_a (workpiece hardness 57 HRC) (with permission to reuse) [59]

2.14 Environmentally Friendly Hard Turning

Whether or not to use coolant is one of the main questions when it comes to the application of hard turning. Hard turning without coolant is the ideal situation and is feasible especially when CBN and ceramic tools are used as these tools can withstand high cutting temperatures. Dry machining eliminates the concerns about the recycle of coolants. Moreover, it is the best option for interrupted cutting because using coolant intensifies the thermal shock while the tool enters or exits the cut. Thermal shock is one of the main reasons for tool breakage. In some applications, using coolant is necessary to control the thermal stability of the workpiece. Application of coolant also provides improved tool life and better surface quality in continuous machining [29]. In such cases, continuous flow of coolant during hard turning must be provided to ensure the coolant reaches the tool tip. High-pressure coolant is one of the most appropriate options to transfer the coolant to the tool tip without being concerned about coolant vaporization due to high temperatures.

Air cooling also seems to be a good environmentally friendly and inexpensive option for hard turning when using CBN tools. Air cooling does not influence surface roughness of the machined surface significantly but significantly reduces the tool wear.

2.15 Hard Milling

Successful implementation of hard turning, especially high-speed turning of hardened steels, generated significant interest in applying the same concept in machining dies and molds. However, dies and molds usually comprise three-dimensional contours and free-form complex surfaces that cannot be achieved by turning. In such applications, milling operation, mainly ball end milling, is employed to accomplish the desired task. Materials from which dies and molds are usually made are commonly categorized as difficult-to-cut materials. This is mainly due to their inherent characteristics such as high strength and hardness, which make them suitable for applications where the material experiences high forces, high temperatures as well as abrasion. D2 tool steels, with hardness exceeding 60 HRC, are a good example of such materials. High-speed machining of die and mold materials offers several advantages over other conventionally machining operations in terms of lower machining cycle time, better dimensional accuracy and surface quality and also prolonged tool life [60–63].

Similar to hard turning, the mechanics of chip formation during milling hardened steels is also dominated by the presence of saw-toothed chips. In addition, due to complex geometry of ball end mill tools, several other mechanisms such as axial and radial thinning are believed to be influential on the mechanics of chip formation [46, 63, 64]. Axial and radial thinning reduce the chip load close to the nose and the end of currently in-cut segment and alter the mechanics of chip formation [63]. This alteration ultimately induces fluctuation to the cutting force and imposes severe dynamic loading on the cutting tool.

2.16 Concluding Remarks

- Hard turning is an economical way to generate high-quality machined surface. The process of hard turning deals with turning of parts with more than 45 HRC hardness. Owing to its higher metal removal rate and the ability to eliminate few production stages, hard turning is considered as an economically feasible alternative for grinding.
- Hard turning is a technology-driven process, as it requires advanced cutting tools, advance rigid machine tools, and also rigid tooling and fixturing systems.
- The most widely used cutting tools for hard turning are CBN and PCBN tools.
- Hard turning can be considered as an environmentally friendly process as it is normally performed in the absence of traditional coolant.
- It is also considered as an environmentally friendly process because it replaces the grinding and the grind wastes.

References

1. Budinski KG, Budinski MK. Engineering materials: properties and selection. Prentice Hall; 2010.
2. Kalpakjian S, Schmid S. Manufacturing processes for engineering materials, vol. 12. 5th ed. Agenda; 2014. p. 1.
3. Suresh R, Basavarajappa S, Gaitonde VN, Samuel G, Davim JP. State-of-the-art research in machinability of hardened steels. Proc Inst Mech Eng Part B J Eng Manuf. 2013;227(2):191–209.
4. Mackenzie DS. History of quenching. Int Heat Treat Surf Eng. 2008;2(2):68–73.
5. Yugandhar T, Krishnan P, Rao CB, Kalidas R. Cryogenic treatment and it's effect on tool steel. In: 6th International Tooling Conference; 2009.
6. Fry T, Willis AM, Steel: a design, cultural and ecological history. Bloomsbury Publishing; 2015.
7. Totten G, Howes M, Inoue T. Handbook of residual stress and deformation in steel. ASM International; 2002.
8. Davies A. The science and practice of welding, vol. 1 & 2. Cambridge; 1984.
9. Campbell FC. Elements of metallurgy and engineering alloys. ASM International; 2008.
10. Krauss G. Deformation and fracture in martensitic carbon steels tempered at low temperatures. Metall Mater Trans A. 2001;32(4):861–77.
11. Roberts GA, Kennedy R, Krauss G. Tool steels. ASM International; 1998.
12. Srinivasan S, Ranganathan S. India's legendary Wootz steel: an advanced material of the ancient world. National Institute of advanced studies; 2004.
13. Srinivasan S, Ranganathan S. Wootz steel: an advanced material of the ancient world. Department of Metallurgy, Indian Institute of Science, Bangalore 1997. http://www.tf.uni-kiel.de/matwis/amat/def_en/articles/wootz_advanced_material/wootz_steel.html

14. Schweitzer PA. Metallic materials: physical, mechanical, and corrosion properties. vol. 19. CRC Press; 2003.
15. Wadsworth J. The evolution of ultrahigh carbon steels—from the great pyramids, to Alexander the great, to Y2k. In: Taleff EM, Syn CK, Lesuer DR, editors. Deformation, processing, and properties of structural materials; 2000. p. 3–24.
16. Sherby O, Oyama T, Kum D, Walser B, Wadsworth J. Ultrahigh carbon steels. JOM. 1985;37 (6):50–6.
17. Sherby O, Wadsworth J. Ultrahigh carbon steels, damascus steels, and superplasticity. Preprint No. UCRL-JC-127180. Istanbul, Turkey: Lawrence Livermore National Library; 1997.
18. Anstis R. Man of iron—man of steel: the lives of David and Robert Mushet. Albion House; 1997.
19. Pye D. Practical nitriding and ferritic nitrocarburizing. ASM International; 2003.
20. Callister WD, Rethwisch DG. Materials science and engineering, vol. 5. New York: Wiley; 2011.
21. Mamlouk MS, Zaniewski JP. Materials for civil and construction engineers. Limited: Pearson Education; 2013.
22. Cardarelli F. Materials handbook: a concise desktop reference. Springer Science & Business Media; 2008.
23. Nakayama K, Arai M, Kanda T. Machining characteristics of hard materials. CIRP Ann Manuf Technol. 1988;37(1):89–92.
24. Kishawy HEA. Chip Formation and Surface Integrity in High Speed Machining of Hardened Steel; 1998.
25. Berns H, Theisen W. Ferrous materials: steel and cast iron. Springer Science & Business Media; 2008.
26. Sharma CP. Engineering materials: properties and applications of metals and alloys. PHI Learning Pvt. Ltd.; 2003.
27. Davis JR. Surface hardening of steels: understanding the basics. ASM International; 2002.
28. Singh M, Ohji T, Asthana R. Green and sustainable manufacturing of advanced material. Elsevier; 2015.
29. Davim JP. Machining of hard materials. Berlin: Springer; 2011.
30. Huddle D. New hard turning tools and techniques offer a cost-effective alternative to grinding. Tool. Prod. Mag. 2001;80:96–103.
31. Soroka D. Hard turning and the machine tool; 2006.
32. Chinchanikar S, Choudhury S. Machining of hardened steel—experimental investigations, performance modeling and cooling techniques: a review. Int J Mach Tools Manuf. 2015;89:95–109.
33. Huang Y, Dawson TG. Tool crater wear depth modeling in CBN HARD TURNING. Wear. 2005;258(9):1455–61.
34. König W, Berktold A, Koch K-F. Turning versus grinding—a comparison of surface integrity aspects and attainable accuracies. CIRP Ann Manuf Technol. 1993;42(1):39–43.
35. Tönshoff H, Arendt C, Amor RB. Cutting of hardened steel. CIRP Ann Manuf Technol. 2000;49(2):547–66.
36. Gaitonde V, Karnik S, Figueira L, Davim JP. Analysis of machinability during hard turning of cold work tool steel (Type: AISI D2). Mater Manuf Processes. 2009;24(12):1373–82.
37. Baránek I. Trends in cutting materials and tools for hard machining. In: Applied mechanics and materials. Trans Tech Publications; 2014
38. Coromant S. Switch to hard-part turning, high-productivity, high-quality finish turning of case-hardened steel surfaces.
39. König W, Klinger M, Link R. Machining hard materials with geometrically defined cutting edges—field of applications and limitations. CIRP Ann Manuf Technol. 1990;39(1):61–4.
40. Kishawy H, Elbestawi M. Effects of process parameters on material side flow during hard turning. Int J Mach Tools Manuf. 1999;39(7):1017–30.

41. Ramesh A, Melkote S, Allard L, Riester L, Watkins T. Analysis of white layers formed in hard turning of AISI 52100 Steel. Mater Sci Eng A. 2005;390(1):88–97.
42. Chou YK, Evans CJ. White layers and thermal modeling of hard turned surfaces. Int J Mach Tools Manuf. 1999;39(12):1863–81.
43. Barbacki A, Kawalec M, Hamrol A. Turning and grinding as a source of microstructural changes in the surface layer of hardened steel. J Mater Process Technol. 2003;133(1):21–5.
44. Griffiths B. Mechanisms of white layer generation with reference to machining and deformation processes. J Tribol. 1987;109(3):525–30.
45. Shaw M, Vyas A. Chip formation in the machining of hardened steel. CIRP Ann Manuf Technol. 1993;42(1):29–33.
46. Elbestawi MA, Srivastava AK, El-Wardany TI. A model for chip formation during machining of hardened steel. CIRP Ann Manuf Technol. 1996;45(1):71–6.
47. Matsumoto Y, Barash M, Liu C. Cutting mechanism during machining of hardened steel. Mater Sci Technol. 1987;3(4):299–305.
48. Chao B, Trigger K. Cutting temperatures and metal-cutting phenomena. Trans ASME. 1951;73(6):771.
49. Huang Y, Liang SY. Effect of cutting conditions on tool performance in CBN hard turning. J Manuf Process. 2005;7(1):10–6.
50. de Oliveira AJ, Diniz AE, Ursolino DJ. Hard turning in continuous and interrupted cut with PCBN and Whisker-reinforced cutting tools. J Mater Process Technol. 2009;209(12):5262–70.
51. Huang Y, Liang SY. Modeling of CBN tool flank wear progression in finish hard turning. Trans Am Soc Mech Eng J Manuf Sci Eng. 2004;126(1):98–106.
52. Lahiff C, Gordon S, Phelan P. PCBN tool wear modes and mechanisms in finish hard turning. Robot Comput Integr Manuf. 2007;23(6):638–44.
53. Diniz AE, de Oliveira AJ. Hard turning of interrupted surfaces using cbn tools. J Mater Process Technol. 2008;195(1):275–81.
54. Dawson TG, Machining hardened steel with polycrystalline cubic boron nitride cutting tools. Georgia Institute of Technology; 2002.
55. Chou YS. Wear mechanisms of cubic boron nitride tools in precision turning of hardened steels; 1994.
56. Davim JP, Figueira L. Machinability evaluation in hard turning of cold work tool steel (D2) with ceramic tools using statistical techniques. Mater Des. 2007;28(4):1186–91.
57. Dawson TG, Kurfess TR. Tool life, wear rates, and surface quality in hard turning. Transactions-north American Manufacturing Research Institution of SME; 2001. p. 175–182.
58. Thiele JD, Melkote SN, Peascoe RA, Watkins TR. Effect of cutting-edge geometry and workpiece hardness on surface residual stresses in finish hard turning of AISI 52100 steel. J Manuf Sci Eng. 2000;122(4):642–9.
59. Thiele JD, Melkote SN. Effect of cutting edge geometry and workpiece hardness on surface generation in the finish hard turning of AISI 52100 steel. J Mater Process Technol. 1999;94 (2):216–26.
60. Kruth J-P, Klewais P. Optimization and dynamic adaptation of the cutter inclination during five-axis milling of sculptured surfaces. CIRP Ann Manuf Technol. 1994;43(1):443–8.
61. Schulz H. High-speed milling of dies and moulds—cutting conditions and technology. CIRP Ann Manuf Technol. 1995;44(1):35–8.
62. Ikeda T. Ultra high speed milling of die steel with ball-nose endmill. In: Proceedings of 2nd ICDMT; 1992. p. 48–56.
63. Kishawy HA, Becze CE. Morphology of chips formed during high speed milling of die and mold tool steel using ball end mills. Technical Papers-Society of Manufacturing Engineers-All Series; 2002.
64. Elbestawi M, Chen L, Becze C, El-Wardany T. High-speed milling of dies and molds in their hardened state. CIRP Ann Manuf Technol. 1997;46(1):57–62.

Chapter 3
Titanium and Titanium Alloys

Abstract Titanium and its alloys are outstanding materials of choice, and their applications are rapidly growing worldwide in high-value markets such as aerospace, marine, power generation, heat exchangers, automotive and biomedical industries. They owe their popularity to their superior characteristics such as high strength to density ratio, also known as high strength to weight ratio or specific strength, as well as high corrosion resistance. Titanium alloys are used in aerospace industry to protect the fuselage, especially in military aircraft, from corrosion and heat damages caused by air friction in supersonic and hypersonic speeds. They are also widely utilized in marine industry to prevent corrosion from seawater or surrounding environment. Although titanium and its alloys are preferable materials by many design engineers, their poor machinability introduces a major drawback that plays a discouraging role in material selection decision. Machining titanium and its alloys requires extra care and attention to machine tool, cutting tool, and cooling strategy as the key elements of each machining system. The main objective is to prevent or minimize vibration, protect the tools from overheating and failure, and also achieve the desired dimensional accuracy and surface quality on the part. This chapter provides the readers with a brief review of the history of titanium, metallurgical aspects of titanium and the effect of alloying elements as well as their mechanical characteristics, and their industrial applications. This chapter also studies titanium and its alloys from machinability prospective in terms of mechanical behavior during machining, mechanics of chip formation, and appropriate cutting tools. The challenges and issues during machining titanium alloys will also be discussed in this chapter.

3.1 Introduction

The rapid development of industries in the twentieth and twenty-first centuries has widened the borders of knowledge and introduced new fields of applications in which parts and products may have never been previously used. Nuclear facilities such as reactors, chemical infrastructures such as oil and gas refineries, energy

H. A. Kishawy and A. Hosseini, *Machining Difficult-to-Cut Materials*, Materials Forming, Machining and Tribology, https://doi.org/10.1007/978-3-319-95966-5_3

transfer lines such as pipelines both onshore and offshore, gas turbine engines, automotive industry, and finally aerospace industry are among many examples of the above-mentioned fields that have been originated or immensely progressed in the twentieth and twenty-first centuries. Although the industrial developments have completely changed and modernized the industry; they have also introduced new challenges and issues that have never been faced in the past. These challenging situations include but are not limited to severe and harsh service condition such as very high, low, or fluctuating temperature, corrosive environments, and dynamic loading. Oil and gas pipelines especially those of offshore, structural components of oil extraction sites, jet engine components such as exhaust nozzles, combustion chambers, compressor, and turbine blades are some examples of parts being used in such harsh conditions.

An ideal material for the above-mentioned applications, where the highest level of safety codes and standards needs to be satisfied, must possess remarkable mechanical and chemical characteristics. Among many preferred characteristics are corrosion resistance, chemical stability or inertness, hot hardness, and low density. Another important feature that is highly desired is high specific strength particularly at elevated temperatures. This characteristic is also known as strength to density ratio or strength to weight ratio. Titanium offers strength to weight ratio higher than those of aluminum and steel. Generally, titanium alloys, depending on their alloying elements and structures, possess all of these characteristics simultaneously. This quality ranks titanium among the best candidates for components used in harsh and demanding service conditions. Moreover, titanium maintains its features at relatively high temperatures, which is a desirable characteristic for industries like power generation and aerospace where working in high temperatures is a common situation. Titanium alloys also demonstrate significant corrosion resistance, which is favorable for architectural components in offshore environments, desalination, nuclear facilities, and oil and gas industries where majority of metals deteriorate and disintegrate gradually when in contact with corrosive chemicals or radioactive materials. Titanium owes its excellent corrosion resistance to the formation of oxide film that protects the surface from corrosion. Furthermore, titanium shows superior elasticity, promising non-magnetic characteristics, and chemical inertness (not at very high temperatures), which make titanium a material of choice for flexible parts, computer industry, and bioimplants applications.

Despite their superior properties both physically and chemically, titanium and its alloys are very expensive due to several challenges encountered during the extraction of titanium from its ore. Titanium alloys are also extremely hard to cut, owing primarily to their mechanical, thermal, and chemical characteristics such as very poor thermal conductivity, high chemical reactivity at high temperatures, low Young's or elasticity modulus, and ultimately their hardening characteristics. Altogether, these features led to encountering several production issues when considering titanium and its alloys for commercial applications. As a result, titanium and its alloys are regarded as difficult-to-cut materials and their applications are relatively limited in the commercial market compared with massive yet quality intensive industries such as military and aerospace where cost is not the prime consideration factor. Despite considerable progress in cutting tool industry and the introduction of harder, more stable, and reliable cutting tools, which revolutionized the machining industry, machining titanium and its alloys still continues to be a problematic task.

3.2 Historical Background and Evolution of Titanium

Titanium (symbol Ti; atomic number 22; and atomic weight 47.9) as an element was initially discovered more than 200 years ago [1], but its worldwide commercial applications were started around the middle of the twentieth century. The employment of titanium opened up entirely new horizons in industry. Owing to its unique characteristics, titanium attracted much attention and was acknowledged as a strategic material in high performance and critical applications such as jet engine components and aircraft fuselage. There is almost an unlimited worldwide supply of raw titanium ore as it constitutes almost 0.6% (0.565%) of the earth's crust [2, 3]. Among the highly abundant structural metals, titanium ranks the fourth after aluminum, iron, and magnesium. Also, among the most abundant elements in the earth's crust, it occupies the ninth place [3, 4]. This amount of natural resources can meet the future market demand generated by further industrial progress. Titanium has two minerals commercially named as ilmenite and rutile. The former is a crystalline iron titanium oxide ($FeTiO_3$) in steel-gray or iron-black color, while the latter is a mineral and mainly consists of titanium dioxide (TiO_2) in a blood red or brownish color [1, 3].

Titanium was initially discovered by the English chemist and mineralogists William Gregor in 1791 in Cornwall (UK) who reported an unidentified element in dark, magnetic iron sand (ilmenite). Later in 1795, when studying rutile, Martin Heinrich Klaproth, a German chemist, reported an oxide of an unidentified element similar to the one reported by Gregor. The term "titanium" was articulated by Klaproth, who decided to use the name titanium after the Titans of Greek Mythology which means the sons of the earth [2].

Despite its vast availability in nature, titanium is yet considered as an expensive material, mainly due to the difficulty of extraction and processing. Titanium is difficult to process and extract from minerals such as rutile; hence, obtaining pure titanium is more expensive than other metals such as steel and aluminum. Scientists and chemists attempted to isolate pure titanium from its ore by means of titanium tetrachloride ($TiCl_4$). However, their attempts to produce titanium with high purity and ductility were obstructed by the high affinity of titanium to make a chemical reaction with oxygen and nitrogen [2]. Several attempts were also made to implement either sodium (Na) or magnesium (Mg) to reduce titanium tetrachloride ($TiCl_4$), but the result was only a small amount of titanium with high brittleness, which was neither economical nor mechanically applicable.

Although titanium dioxide was extracted from sand and rock samples by William Gregor and Martin Heinrich Klaproth in the late eighteenth century, it was not utilized to its full potential until more than a century later. In 1910, a pure sample of titanium was extracted in USA by a chemist named Matthew Hunter. In a laboratory setup, he reacted the sodium (Na) with titanium tetrachloride ($TiCl_4$) at 800 °C under excessive pressure to isolate titanium [5]. The process invented by Hunter was named after him and is currently known as Hunter process. A handful of different techniques were introduced after Hunter's process to obtain pure titanium, but none was successful or economically feasible.

The endless efforts made by the scientists to find an economically feasible approach to extract titanium from its ore finally paid off in the twentieth century when William Kroll, a Luxembourgish metallurgist, extended Hunter's process and developed a new multi-step technique (1937–1940) to extract titanium from its ore. The process introduced by Kroll uses magnesium to reduce titanium tetrachloride ($TiCl_4$) in the presence of an inert gas. The output of this process is a porous and spongy metal which is called "titanium sponge" [2]. Following this step, an electric arc vacuum furnace melts the titanium into ingots. The vacuum furnace prevents non-inert gasses from being mixed with the titanium during the melting process. Finally, the ingots are converted to billets or rods as the final products. Due to the high reactivity of titanium with other metal or non-inert gases at high temperatures, extra attention must be paid during the refining process in order to prevent contamination or reaction with nitrogen and oxygen [6]. The Kroll process is very efficient in comparison with the previously developed methods to extract titanium; hence, it is still the dominant process for the extraction of titanium and is widely used in industry. Although Kroll's process improved the feasibility of titanium extraction, titanium is still a very expensive material because the Kroll process is not a continuous process. It involves multiple phases each of which must be completed at high temperatures for a batch of ore that makes the process still labor intensive and costly [4].

Another major breakthrough in the history of titanium was made after World War II. In the late 1940s and early 1950s, the USA and the Soviet Union, whom by that time were the pioneers in aerospace industry, faced a new barrier toward their tremendous effort in gaining military advantage and air superiority. The challenge was to produce aircraft capable of flying high and fast to avoid enemy threat and increase the chance of survival during reconnaissance, interception, or combat missions. Achieving higher speeds required powerful engines capable of delivering enough thrust as well as strong yet lightweight fuselage to sustain high g-forces that could be encountered during combat maneuvers. Higher thrust is typically accompanied by higher temperatures that only a few materials at that time were able to sustain; yet none of them was lightweight enough for the application.

The increasing demands for a material with superior characteristics drew the attention to titanium and its alloys and encouraged the US government to financially support the construction of large capacity titanium sponge production plants [2]. The American government also stockpiled titanium ore for the sole purpose of enhancing its military operations. At the same time, in 1950, the Soviet Union began engineering the use of titanium for military and submarine applications.

3.3 Metallurgy of Titanium

Titanium is a transition metal that offers lightweight and temperature resistance simultaneously owing to its low density and a high melting point. A combination of these two features made titanium an attractive material for several industrial applications. Nevertheless, the characteristics of titanium alloys mainly depend on microstructure and chemical composition. Titanium exhibits two different crystal structures: alpha (α) and beta (β). These two crystal structures are split at 882 °C as alpha phase (α) with hexagonal close-packed (HCP) metallic crystal structure undergoes allotropic transformation and is converted to beta phase (β) with body-centered cubic (BCC) when temperature exceeds the 882 °C threshold and below the melting point of 1668 °C [7–9]. This borderline temperature is for pure titanium and can be altered by alloying it with other elements [9]. Between the two forms of crystal structures, the alpha phase with body-centered cubic structure has higher ductility than the beta phase with hexagonal close-packed structure. This is mainly due to the higher deformation capability of BCC structure than that of HCP. Hexagonal close-packed structure has limited deformation capability and hence reduces the ductility of the beta phase [10]. Figure 3.1 demonstrates the two possible crystal structures of titanium.

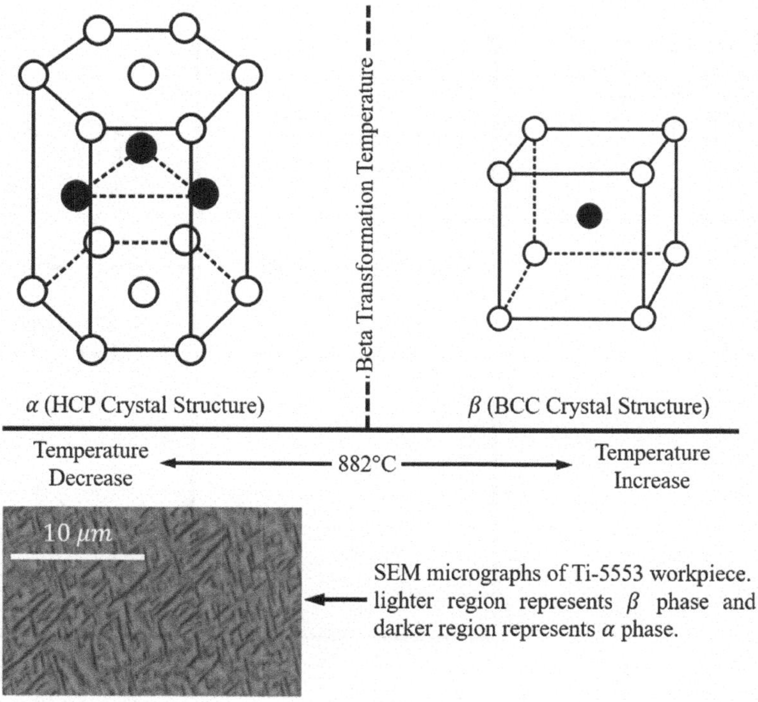

Fig. 3.1 Two allotropic forms of titanium

Titanium in its pure form demonstrates adequate corrosion resistance; however, its physical and mechanical characteristics can be widely altered by adding alloying elements [9] to obtain different titanium alloys. Alloying elements either increase or decrease the transformation temperature; thus, based on their effect, they can be divided into α-stabilizers and β-stabilizers [10].

The alpha phase can be stabilized by adding elements such as aluminum (Al), gallium (Ga), oxygen (O), nitrogen (N), and carbon (C) with aluminum being a very popular choice due to its low density and strengthening effects [7]. These elements raise the transformation temperature and allow alpha phase to stay stable even at higher temperatures; thus, they are known as α-stabilizers. For instance, titanium–aluminum alloys exhibit good tensile strength and creep stability in a range of temperatures varying from room temperature up to 300 °C [7]. Due to their high solubility in alpha and beta phases, adding elements such as tin and zirconium does not substantially affect the transformation temperature; however, the addition of these elements strengthens the α phases [7].

The transformation temperature can be reduced by adding vanadium (V), molybdenum (Mo), niobium (Nb), iron (Fe), chromium (Cr), nickel (Ni), manganese (Mn), and cobalt (Co) which are known as β-stabilizers [8, 9]. β-stabilizers are divided into β eutectoid and β isomorphous. Among the above-mentioned β-stabilizers, molybdenum, vanadium, and niobium are common examples of

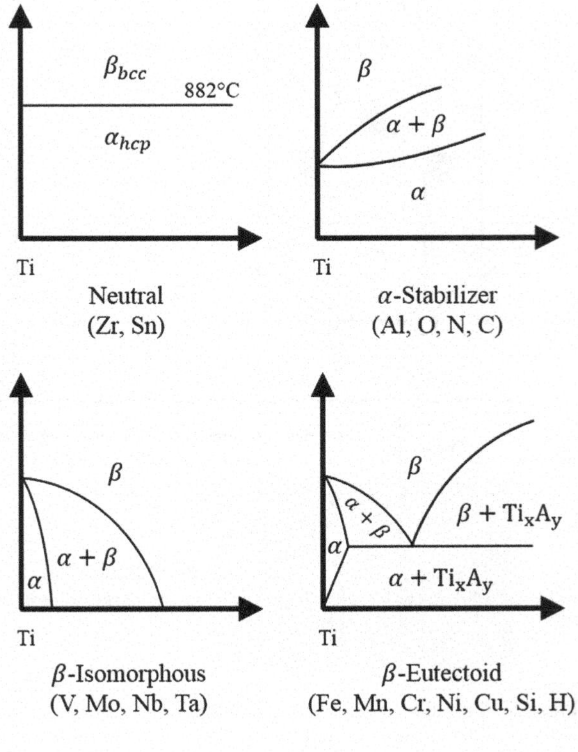

Fig. 3.2 Phase diagrams of titanium under the effects of alloying elements (with permission to reuse) [2, 10]

β-isomorphous elements. Figure 3.2 demonstrates the effects of alloying elements on the phase diagram of titanium.

Although more than 100 titanium alloys can be produced by mixing titanium with other alloying elements, only a fraction of them (20–30) is commercially used nowadays [10]. Titanium alloys are usually categorized, based on their crystallographic phases at room temperature [11, 12], into the following main categories:

- Alpha (α) alloys
- Alpha-Beta (α + β) alloys
- Beta (β) alloys.

With further subdivision into

- Metastable beta (β) alloys
- Near-alpha (α) alloys.

A detailed description of each group and subgroup of titanium alloys is presented in the following sections.

3.3.1 Alpha (α) Alloys

The alpha alloys which are also known as commercially pure (cp) titanium are single phase alloys [13]; however, they may contain some amount of α-stabilizer contents [7, 8]. Alpha alloys typically retain their tensile strength up to 300 °C. Moreover, they show good oxidation and creep resistance, reasonable ductility, great weldability, and relative high toughness at cryogenic temperatures. They are mostly employed in chemical processing industry where formability and corrosion resistance are the key design factors. They are also used in cryogenic applications such as production of storage tanks for liquid hydrogen [14]. Among the α-stabilizers, oxygen improves strength while simultaneously reducing ductility. However, as alpha alloys consist of a single phase, their mechanical properties cannot be improved by heat treatment; hence, they are relatively weaker than other titanium alloys and demonstrate lower strength [13]. The members of this group of titanium alloys show strength ranging from 240 to 740 MPa at room temperature [10].

3.3.2 Near-Alpha (α) Alloys

As stated before, beta crystal structure is not stable at room temperature and exists in temperatures above 882 °C; however, it has been shown that [8] adding a controlled amounts of beta stabilizers (1–2%) helps the formation of beta phase even below the transformation temperature down to room temperature. This category of titanium alloys, which includes high amounts of α-stabilizers and small amount of β-stabilizers, is usually referred to as near-alpha alloys. Classically known as high-temperature alloys, near-alpha alloys combine the desired characteristics of both alpha alloys and alpha-beta alloys. They exhibit exceptional creep resistance similar to that of alpha alloys while concurrently possessing high strength of alpha-beta alloys. They are capable of maintaining their mechanical properties even in the range of 400–520 °C [9], which is higher than the service range of alpha alloys. This feature makes them a material of choice to replace heavier nickel-based superalloys in aeroengine component wherever applicable [15]. Examples of alloys that are near the alpha phase include IMI 685/Ti–6Al–5Zr–0.5Mo–0.25Si and Ti 8–1–1/Ti–8Al–1Mo–1V. It must be noted that among different possible microstructures that can be developed by alloying titanium, the best corrosion resistance and highest weldability are offered by pure titanium, alpha alloys, and near-alpha alloys [16].

3.3.3 Alpha-Beta (α + β) Alloys

Alpha-beta alloys are an amalgamation of α- and β-stabilizers, and their microstructure is a combination of both alpha and beta phases. This family of titanium alloys can be obtained by adding more β-stabilizer (4–6%) to the base metal [8]. Altering the amount of beta stabilizer generates different microstructures that eventually lead to a wide range of mechanical characteristics. Mechanical properties of this family of titanium alloys can be summarized as follows:

- Heat treatable [8]
- Excellent ductility [13]
- Medium to high strength in a range of temperature from 350 to 400 °C [7]
- Creep strength at high temperatures (lower than that of alpha alloys).

One of the most well-known members of this family, which is by far the most widely used titanium alloys, is Ti 6–4 (Ti–6Al–4V) [7]. More than half of all titanium alloys, which are currently used in different industrial sectors, are of this composition.

3.3.4 Metastable Beta (β) Alloys

If the β-stabilizer content is increased more than those of alpha and alpha-beta alloys up to 10–15%, at room temperature the beta phase will have a metastable state [8]. Metastable beta alloys exhibit great capability to be forged or hardened. Under controlled alloying conditions, their strength can reach as high as 1400 MPa [10]. If properly heat treated, they also show great strength and excellent fracture toughness [17] within a wider temperature range, which makes them a primary option for structural components in aerospace industry. The complex microstructure of this family allows the metallurgist to tailor the material properties for high toughness and high strength.

3.3.5 Beta (β) Alloys

If the amount of β-stabilizer is large (30%), the resultant titanium alloy is called beta alloy. β-stabilizers increase hardenability and density; hence, beta alloys have high density. Beta alloys suffer from lack of ductility, but they exhibit good creep resistance; thus, they are good candidates for the burn and corrosion resistance applications [8]. At room temperature, beta alloys show almost equivalent strength to alpha-beta alloys. However, at an increase in temperature, alpha-beta alloys are stronger than beta alloys.

Generally, among titanium alloys, alpha alloys display higher beta transition temperature, higher modulus of elasticity, better machinability, greater weldability, higher high-temperature strength, and finally better creep and heat resistance. Alternatively, beta alloys exhibit greater specific density, room temperature strength and toughness, age hardenability, plastic and superplastic formability, heat treatability, and strain-rate sensitivity. Alloys in between are engineered for desired applications by enhancing specific aspects. Table 3.1 summarizes the variation of properties among different titanium alloys. It should be noted that TiNi, Ti_3Al (α2), and TiAl (γ) intermetallic-type alloys are not included in this table.

Table 3.2 shows some of the important titanium alloys that are commercially used in industry.

3.3.6 Titanium Aluminides

Titanium aluminides are interesting members of the titanium family. They are classified as intermetallic chemical compounds in which ordered alloy phase is formed between two metallic elements [19–21]. Due to this special form of microstructure, titanium aluminides demonstrate great characteristics such as high-temperature corrosion resistance, strength, and stiffness [19]. Some titanium aluminides such as Ti_3Al (α2) and TiAl (γ) can be potentially implemented at temperatures around 650 and 800 °C, respectively [10]. Titanium aluminides also have low density and excellent creep-rupture resistance [18]. They owe their outstanding creep resistance to their ordered structure; however, their microstructure imposes some limitations like reduced dislocation mobility which leads to very low ductility and fracture toughness [8, 19].

Table 3.1 Effects of major alloying elements on the properties of titanium alloys (with permission to reuse) [18]

Major families of titanium alloys		
α and near α	α + β	β and near β
Commercially pure Ti	Ti–5Al–2.5Fe	Ti–3Al–8V–6Cr–4Mo–4Zr
Ti–5Al–2.5Sn	Ti–5Al–2Mo–2Fe	Ti–4.5Al–3V–2Mo–2Fe
Ti–5Al–6Sn–2Zr–lMo	Ti–5Al–3Mo–4Zr	Ti–5Al–2Sn–2Zr–4Mo–4Cr
Ti–6Al–2Sn–4Zr–2Mo	Ti–5Al–2.5Fe	Ti–6Al–6Fe–3Al
Ti–8Al–lMo–l V	Ti–6Al–7Nb	Ti–10V–2Fe–3Al
	Ti–6Al–4V	Ti–13V–11Cr–3Al
	Ti–6Al–6V–2Sn	Ti–15V–3Cr–3Al–3Sn
	Ti–6Al–2Sn–4Zr–6Mo	Ti–35V–15Cr
		Ti–8Mo–8V–2Fe–3Sn
		Ti–11.5Mo–6Zr–4.5Sn
		Ti–30Mo, Ti–40Mo
		Ti–13Nb–13Zr
		Ti–25Pd–5Cr
		Ti–20Cr–0.2Sn
		Ti–30Ta

Variation of characteristics from α and near α to β and near β titanium alloys		
β-Transus temperature	→	Decreasing
Specific density	→	Increasing
Room temperature strength	→	Increasing
Room temperature toughness	→	Increasing
Modulus of elasticity	→	Decreasing
Machinability	→	Decreasing
Age hardenability	→	Increasing
Heat resistance	→	Decreasing
Weldability	→	Decreasing
High-temperature strength	→	Decreasing
Heat treatability	→	Increasing
Plastic formability	→	Increasing
Strain-rate sensitivity	→	Increasing
Superplastic formability	→	Increasing
Creep resistance	→	Decreasing

3.4 Characteristics of Titanium and Its Alloys

Titanium and its alloys have gained widespread use in different industrial sectors due to their favorable properties such as high strength to weight ratio, low thermal conductivity, and resistance to wear. When compared to most other commonly used materials in industrial applications such as iron, nickel, and aluminum, titanium

Table 3.2 Important commercial titanium alloys (with permission to reuse) [2]

Common name	Alloy composition (wt%)	T_β (°C)
α alloys and CP titanium		
Grade 1	CP–Ti (0.2Fe, 0.18O)	890
Grade 2	CP–Ti (0.3Fe, 0.25O)	915
Grade 3	CP–Ti (0.3Fe, 0.35O)	920
Grade 4	CP–Ti (0.5Fe, 0.40O)	950
Grade 7	Ti–0.2Pd	915
Grade 12	Ti–0.3Mo–0.8Ni	880
Ti–5–2.5	Ti–5A1–2.5Sn	1040
Ti–3–2.5	Ti–3A1–2.5V	935
α + β alloys		
Ti–811	Ti–8A1–1V–1Mo	1040
IMI 685	Ti–6A1–5Zr–0.5Mo–0.25Si	1020
IMI 834	Ti–5.8A1–4Sn–3.5Zr–0.5Mo–0.7Nb–0.35Si–0.06C	1045
Ti–6242	Ti–6A1–2Sn–4Zr–2Mo–0.1Si	995
Ti–6–4	Ti–6A1–4V (0.20O)	995
Ti–6–4ELI	Ti–6A1–4V (0.13O)	975
Ti–662	Ti–6A1–6V–2Sn	945
IMI 550	Ti–4A1–2Sn–4Mo–0.5Si	975
β alloys		
Ti–6246	Ti–6A1–2Sn–4Zr–6Mo	940
Ti–17	Ti–5A1–2Sn–2Zr–4Mo–4Cr	890
SP–700	Ti–4.5AI–3V–2Mo–2Fe	900
Beta-CEZ	Ti–5A1–2Sn–2Cr–4Mo–4Zr–lFe	890
Ti–10–2–3	Ti–10V–2Fe–3A1	800
Beta 21S	Ti–15Mo–2.7Nb–3Al–0.2Si	810
Ti–LCB	Ti–4.5Fe–6.8Mo–1.5Al	810
Ti–15–3	Ti–l5V–3Cr–3Al–3Sn	760
Beta C	Ti–3A1–8V–6Cr–4Mo–4Zr	730
B120VCA	Ti–13V–11Cr–3A1	700

stands out with the highest strength to density ratio. Titanium's strength to density ratio is approximately 222.22 MPa cm^3/g compared to iron's ratio of 126.58 MPa cm^3/g, nickel's ratio of 112.36 MPa cm^3/g, and aluminum's ratio of 185.19 MPa cm^3/g [2]. Among metals that are commonly used in industry, only the highest strength steels exceed titanium in terms of strength to weight ratio. Titanium alloys offer a wide range of yield strength from 800 to 1200 MPa. Exceptional yield and tensile strength of 1590 and 1620 MPa can also be obtained by double aging treatment of a certain titanium β alloy called TIMETAL 125 (Ti–6V–6Mo–6Fe–3Al) [10].

The high strength to weight ratio is a desired property factor especially when minimizing the weight and concurrently maintaining the structural integrity is of prime important. This allows aerospace engineers to design aircraft that fly faster and are more fuel-efficient. The high strength to density ratio is directly related to the crystal structure of titanium, which is a hexagonal close-packed (HCP) structure in room temperature. This form of crystal structure possesses a high atomic packing factor (APF) of approximately 0.74. Such a crystal structure retains a high bond density, which makes it difficult for dislocations to flow through the material when load is applied. This consequently leads to higher material's strength [22]. Among the commonly used materials in industry, only nickel and aluminum possess a similar APF of 0.74 owing to their face-centered cubic (FCC) structure. This does not necessarily result in equivalent strength, since the strength of the metallic bonds between nickel atoms and aluminum atoms is much weaker than the metallic bonds between titanium atoms [22].

Titanium also possesses excellent thermal characteristics including high melting point and low thermal conductivity. Such characteristics qualified titanium as a candidate in applications where the components are exposed to high temperatures. High melting point affects the high-temperature mechanical properties of the material. This is mainly due to the fact that the temperature at which the crystal structure changes (transition temperature) is a percentage of melting point. Therefore, as the melting point of the material increases, typically the transition temperature increases as well. Therefore, the material maintains its mechanical properties even at higher temperatures.

As previously mentioned, at 882 °C the crystal structure of titanium transforms from a hexagonal close-packed (HCP) to a body-centered cubic structure (BCC) [9]. The change in crystal structure greatly affects the mechanical properties as BCC structure has a lower APF of about 0.68 in comparison with APF of 0.74 for HCP. Decreasing the APF lowers the quantity of metallic bonds and consequently reduces the modulus of elasticity (E) and the shear modulus (G) of the material; see Fig. 3.3 [2]. As illustrated in Fig. 3.3, Young's modulus dramatically decreases from the initial value of 115 GPa at room temperature to approximately 58 GPa in the vicinity of transition temperature of 882 °C ($\sim$1155 K). The shear modulus G exhibits almost the same trend and drops from about 42 to 20 GPa over the same temperature range [2]. Moreover, at higher temperatures, titanium becomes more and more reactive to oxygen, which promotes the formation of very brittle titanium oxide.

Another important thermal property of titanium and its alloys is low thermal conductivity that makes it difficult for the heat to propagate through the material. This property can be favorable in the case of local heating where only a specific area of the material is subjected to heat. In such circumstances, the material properties in the heat-exposed zone may decrease, but the rest of the material would be capable of maintaining its characteristics. On the other hand, low thermal conductivity can also be a challenge specially during machining of titanium and titanium alloys. During machining, the localized heat due to plastic deformation and friction adversely affects the tool material and rapidly deteriorates the cutting tool. Moreover, the speed of deformation will be much faster than that of heat

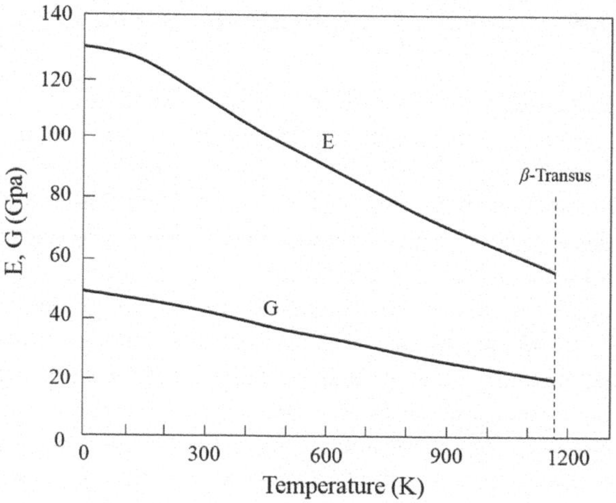

Fig. 3.3 Modulus of elasticity *E* and shear modulus *G* as a function of temperature of α-titanium polycrystals (with permission to reuse) [2]

propagation in the cutting zone, which promotes adiabatic shear and local deformation. This topic will be discussed in detail later in this chapter.

Pure titanium is very reactive to oxygen, especially at elevated temperatures. When pure titanium or its alloys are exposed to the atmosphere, the outer surface reacts with the air and produces a thin layer of titanium dioxide called titania, which is very hard and brittle. This oxide layer possesses mechanical properties different from pure titanium. The formation of oxide can cause problems if the material is used at a temperature above 600 °C. Above this temperature, titanium becomes highly reactive and diffusion of atmospheric oxygen through the oxide layer makes the material very brittle which will ultimately lead to the failure of the part [2].

Formation of the oxide layer can be advantageous as it acts as a barrier, which shields the material against undergoing any chemical reactions in low temperatures. The extra resistance to chemical reactions, due to the protective oxide layer, protects the material from corrosion. For this reason, titanium and its alloys are considered to be very biocompatible and are widely used for surgical implants as the material is capable of resisting corrosion from the organic body fluids [23]. Another reason why titanium and its alloys are widely used throughout the medical field for implants is due to the high dielectric constant of titania. The high dielectric constant allows the material to bond more easily to organic tissue such as bones. This characteristic gives titanium a large advantage over most other surgical implants as titanium is capable of osseointegration, meaning it is capable of bonding to the bones without the need for adding an adhesive. This not only helps increasing the life of the implant, but also strengthens the bond between the bone and the implant material [23].

3.5 Industrial Applications of Titanium and Its Alloys

Industries that operate in the most demanding environments, under extreme loads, and in the presence of heavily corrosive agents such as oil and gas, largely depend on titanium and its alloys for placement in rigorous situations. Titanium and its associated alloys are well known for their high strength to weight (strength to density) ratio, which can provide vast weight savings compared to the utilization of traditional steel. In addition to its strength, titanium also resists creeping in high-temperature environments, as well as possessing high corrosion resistance. Titanium and its alloys are high-quality materials that come with high costs, but the significance of their positive impact on the industries usually compensates their high price, especially in demanding applications where high safety and reliability are key design considerations. All of these properties combined together make titanium and its alloy attractive materials for use in many industrial sectors including aerospace, chemical, biomedical, sport, and many others. As a result, titanium will be a metal of choice for various applications for many years to come.

3.5.1 Aerospace Applications

Although the high costs associated with processing titanium hinder its economic feasibility, in the multi-billion-dollar aerospace industry where safety and performance are the prime concerns, titanium is the most suitable material for many applications. Weight savings are paramount when it comes to taking flight, and titanium has a superior strength to weight ratio compared to aluminum and steel as its traditional counterparts [24]. This allows manufacturers to reduce the overall weight of the aircraft, use the savings to minimize expenses such as fuel, or reinvest that weight in other areas like better passenger accommodations or larger fuel capacity to increase the flight range. Developed by Lockheed in 1960, strategic reconnaissance SR-71 known as Blackbird was the first military aircraft with an airframe that contains 93% titanium alloys. SR-71 was able to fly at Mach 3 + (three times faster than the speed of sound). Flying at such a high speed exposed SR-71 to temperatures as high as 232 °C at its aft midsection rising up to 510 °C near the engine exhaust [3, 25] (Fig. 3.4).

Boeing and Airbus are two major aerospace companies with many of their aircraft models taking advantage of titanium alloys for the landing gear assemblies. Not only can the landing gear support the weight of the aircraft above it, but it also effectively translates the huge impacts upon landing into the aircraft's shock absorption system. By simply reducing the weight of the landing gear, there are many other subordinate benefits as well. Lighter landing gear means smaller hydraulic components to retract and deploy the assembly and less of a load on the plane's electrical-hydraulic system, again helping to reduce the overall weight of the aircraft or have the option to relocate that weight elsewhere. In planes, a large

Fig. 3.4 SR-71 Blackbird with airframe comprises 93% titanium [26]

portion of the landing gear system is composed of a high-strength beta (β) alloy of titanium.

Another obstacle that aerospace engineers must overcome when designing an aircraft's components is selecting proper corrosion-resistant materials. Like aluminum and its alloys, titanium also forms a strong oxidizing layer when it is exposed to air. As stated earlier, the layer of titanium dioxide is very resilient and protects the remaining titanium from any further oxidation and deterioration in strength. Referring back to the use of titanium in landing gear assemblies, its resilience to corrosion is very important for landing in areas near ocean spray and winter areas that use salt on runways and taxiways. Especially with a steel landing gear structure, salt would exponentially increase the rate of decay in the steel structure as seen on many Canadian cars where roadways are often covered with salt. A failure in any part of the landing gear assembly could be catastrophic, and the use of a titanium alloy lowers the risk of a part failure. In addition, when titanium alloys are employed to build the vanes within the aircraft's engine, the protective properties of titanium's oxidized layer vastly outperform aluminum or steel alloys. Despite its protective layer of oxide, aluminum will eventually develop corrosion pits and have a greatly shortened service life. These pits can then cause increased stresses on the part and even fatigue cracks. In the worst-case scenario, a vane from the engine's compressor could totally fail and break apart into the engine, resulting in a disastrous event. Metal debris could pierce the engines surrounding covers and possibly even penetrate into the fuselage of the aircraft resulting in a loss of cabin pressure. Furthermore, the pieces of the broken vane would continue to travel through the engine, destroy many more compressor and turbine vanes, eventually destroy the engine, and render it useless. During inclement weather, a catastrophic failure of that magnitude may even result in a plane crash and the loss of human life.

Smaller parts onboard the aircraft can also be produced by using titanium. These parts are typically within the hydraulic system but more importantly inside the aircraft's engines. In a jet engine, all of the titanium's benefits truly become prominent. Their high tensile strength to weight ratio is particularly appealing to aerospace engineers when designing the compressor and turbine vanes. Obviously, the main intention of a compressor or turbine blade is to take rotational shaft energy and translate the energy into the air (or vice versa), but if the vane is excessively heavy, a lot of energy is wasted when speeding up or slowing down the rotational mass. To minimize wasted energy when decelerating or accelerating the engine's vanes, the weight of the vanes themselves needs to be minimized by utilizing titanium alloys. Improved throttle response and maneuverability are also possible with a lesser rotational mass since the engine will be able to alter its speed more rapidly. The reduced mass also lowers fuel consumption and increases flight range.

Titanium is also used to build several other components in the jet engines as shown in Fig. 3.5. The components of jet engines, particularly those that are located in the proximity of where the fuel is burned, experience extremely high temperatures during their service life. For instance, fan case, compressor vanes, and compressor blades [27] undergo the majority of the stress through application as they are the moving parts. Such components must be able to withstand these high temperatures while maintaining their functionality. Titanium is the material of choice for such applications due to its low thermal conductivity, low thermal expansion, and temperature resistance.

In addition to the components shown in Fig. 3.5, another potential application for the use of titanium alloys in aerospace industry is in building the aircraft turbines. The use of titanium in aircraft turbines is very recent, especially for civilian aircraft. It was not until 2011 when the first commercially viable engines were produced using titanium–aluminum alloys in the low-pressure turbine section [28]. Ti–Al alloys exhibit excellent oxidative resistance, high-temperature creep strength, and low density, which make them ideal to be used in aircraft turbines. The use of Ti–Al in the engines contributed to the GEnx engines being able to drop their fuel consumption, NOx emissions, and operation noise by 20, 80, 50%, respectively, compared to similar engines [28].

While it is apparent that titanium and its alloys are often used in components around the aircraft, one might consider using titanium as a member of the aircraft's structure. The midsection of an airplane where the wings meet the fuselage and where the rear landing gear is stored does not have much excessive space remaining for the support structure. The aforementioned section of the plane also undergoes huge stresses during takeoff and landing because all of the lift generated by the aircraft's wings must be translated into the aircraft body at that same point. This means that the support structure underlying the midsection of the cabin, connecting the wings to the fuselage, needs to be extremely strong but has minimal weight or

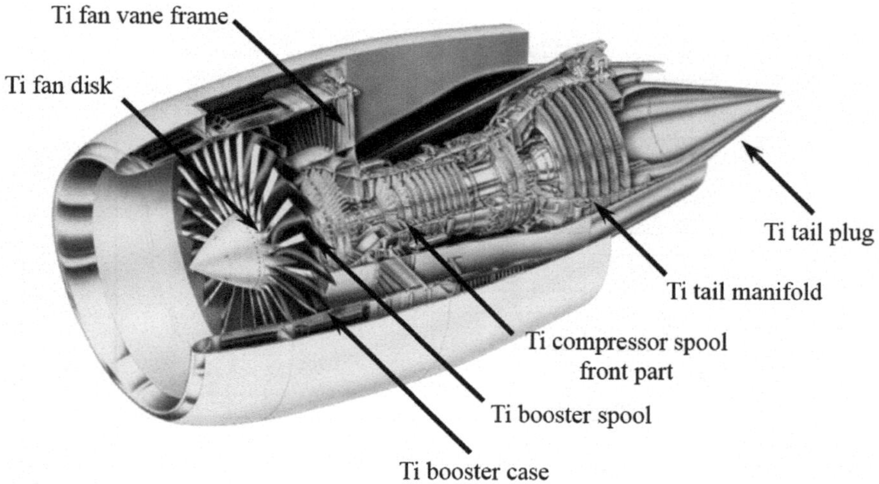

Fig. 3.5 Titanium usage in the GE-90 aeroengine (with permission to reuse) [2]

size footprint. Traditionally, aluminum is the material of choice, but with the overall growing size of aircraft the size of the beam needed to support the anticipated weight is simply too large for the space available. Steel is another option for the job, but at the expense of a much higher weight addition to the aircraft. Thus, titanium once again is the ideal candidate for the job. Its only main downside is the costs associated with the raw material and fees to process it. Figure 3.6 demonstrates the demand for titanium in commercial aerostructures in the period between 2010 and 2016 [27].

During flight, planes experience cryogenic temperatures, in which the material of the airframe must be able to withstand. Titanium possesses this ability, as it is resistant to embrittlement at cryogenic temperatures. As a result, unalloyed titanium sheets are usually used in the construction of the airframes.

3.5.2 Chemical and Petrochemical Applications

Chemical and petrochemical industries extensively employ titanium and its alloys because of their ideal mechanical and chemical properties. Titanium is among the most readily available corrosion-resistant materials in the world, and thus it is an optimal candidate for parts and equipment to be used in chemical processing industry. The traditional materials of choice for such applications were stainless steels and other nonferrous metals. Since the process of producing chemical products generally involves high temperatures in the presence of chemical reactants, traditionally used materials are not capable of sustaining over long periods. Short

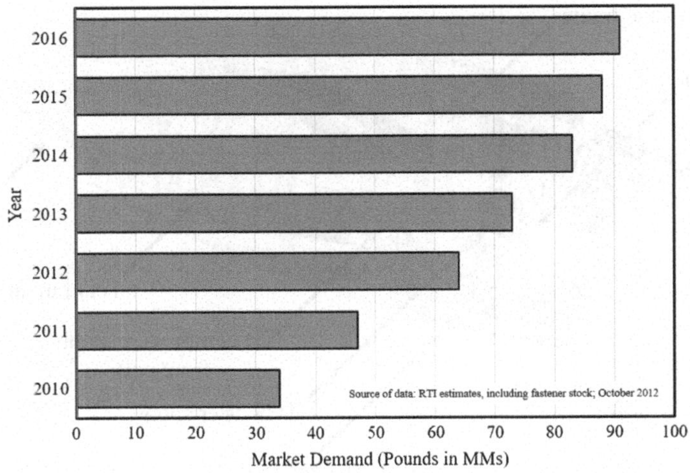

Fig. 3.6 Commercial aerostructure titanium demand [27]

life span means frequent maintenance or replacement, which in turn jeopardizes the economic feasibility of the process.

Due to their high thermal capacity and corrosion resistance, titanium and its alloys expand components' life span, reduce cost, and improve quality; hence, they have made a huge impact on the petrochemical industry. One example is the production of crude oil with increased sulfur and carbon dioxide content. This has driven the need for chemical processing to change over to titanium [10] to elongate the equipment life span in the presence of higher chemical content and high corrosion levels. In terms of the life cycle analysis [10], titanium has proven to be a better choice in the long-term production of both chemical and petrochemical products. Although there is much higher cost for the initial production of equipment due to the cost in producing and refining titanium, the high initial investment will eventually be paid off over the longer life span that can be achieved.

3.5.3 Automotive Applications

Automotive industry is another field of applications for titanium and its alloys [2, 29]. Components in automobiles are designed from titanium due to its high fatigue strength, low density, and high-temperature resistance. Automotive manufacturers are willing to pay for the higher cost of titanium if there are significant reductions in rotating and oscillating masses within the vehicle. The reduction in moving masses thereby reduces vibrations within a vehicle and thus reduces the need for damping systems and materials. One use of titanium is to construct the piston connecting rods in sports cars. However, only approximately 33% of the connecting rod mass contributes to oscillating mass, which limits the resulting gains of creating connecting rods out of titanium [29]. Another application of titanium alloys in automotive industry is in the engine valve train. The valve faces are subjected to high-temperature changes due to igniting gases. Titanium's low density and high-temperature resistance allow for the creation of durable yet lightweight valves that can increase engine performance. Another example of using titanium alloys in automotive industry is the valve springs. Making the springs out of titanium significantly reduces the weight and leads to significant weight reduction in the oscillating mass of the engine. The third example is the valve cups where the main benefit is again weight reduction, but a surface coating is required to enhance wear resistance. A fourth area to apply titanium in automobiles is in the manufacturing of sealing washers. The main reason for this application is the high corrosion resistance of titanium. Other reasons include weight reduction and the ability of titanium to deform to create a high-quality seal. In summary, the main purpose of titanium in the automotive industry is to minimize vibrations by reducing the mass of moving components. This task is accomplished by manufacturing components that can take advantage of other properties of titanium.

Table 3.3 summarizes the typical applications of some common titanium alloys [30, 31].

Table 3.3 Typical application of common titanium alloys in industry

Alloy	Reason for use and application
Unalloyed Ti Grade 1	Weldable, high ductility during fabrication and low strength in service, low iron and interstitial content, and high ease of formability. Used in medical implants
Ti–5Al–2.5Sn	Retains toughness and ductility in the annealed condition and is used in cryogenic applications, typically in aerospace applications
Ti–8Al–1Mo–1V	High weldability, high strength, and low density, typically used for fan blades
Ti–6Al–2Sn–4Zr–2Mo	High strength and toughness, high creep resistance, and stable at relatively high temperatures used in forging and flat rolling
Ti–2.25Al–11Sn–5Zr–2Mo	High-temperature creep strength and stability, as well as high strength in short application periods, typically used in jet blades, jet wheels, and bulkhead forging
Ti–5.8Al–4Sn–3.5Zr–0.7Nb–0.5Mo–0.35Si	Highest service temperature of titanium alloys, typically used in fan and compressor blades in aerospace engine cycles
Ti–6Al–4V	Most commonly used alloy with a mediocre strength and resistance to temperature, a wide range of use from aerospace components to prosthetic implants
Ti–4Al–4Mo–2Sn–0.5Si	Medium-strength alloy commonly used in aerospace components

(continued)

Table 3.3 (continued)

Alloy	Reason for use and application
Ti–6Al–6V–2Sn	High strength at relatively high temperatures, typically used in aerospace components although use is limited due to unfavorable material properties
Ti–6Al–2Sn–4Zr–6Mo	Commonly used for compressors or fan blades which undergo large loads over a long period at elevated temperatures
Ti–10V–2Fe–3Al	Highest combination of strength, any toughness as compared to other alloys, used in application requires uniform properties at the surface of the material
Ti–3Al–8V–6Cr–4Mo–4Zr	Most commonly used in fasteners, torsion bars, and springs among other applications
Ti–15Mo–3Al–2.7Nb–0.2Si	One of the most important titanium alloys due to its high oxidation resistance as well as its creep resistance; also, notably the alloy is resistant to the working fluids used in commercial aircraft exhaust systems
Ti–15V–3Cr–3Al–3Sn	Used in applications requiring high strength in low-temperature applications

3.6 Challenges in the Machining of Titanium and Its Alloys

It has already been a well-recognized fact that among the advanced materials and alloys that currently experience rapid growth in their application, titanium is one of the fastest growing ones. Its wide application in different industrial sectors will continue to grow due to the market trend toward lighter yet safer materials, more specifically in aerospace industry. As previously mentioned in this chapter, titanium owes its reputation to characteristics such as relative high-strength level for a given low mass, great corrosion resistance, and relatively good resistance against high temperature. These properties can also be customized and tailored for any specific application by alloying titanium with other elements.

However, the above-mentioned characteristics that make titanium an interesting choice for several applications also draw a barrier toward its widespread use from the manufacturing perspective. In addition to its expensive price, titanium alloys suffer from poor machinability that makes them very difficult to cut. Hence, the machine tools used for machining titanium alloys must be made of much stronger materials to prevent vibration. Also, tool life is expected to be very short. These contradictory features (good mechanical characteristics versus high price and poor machinability) encourage the tool manufacturer, with decades of experience in producing cutting tools for machining industries, to invest capital funds in research and developments (R&D). They are also urged to closely work with researchers in universities and research centers as well as material suppliers to study the machinability of titanium and its alloys. The main objective of these research efforts is to acquire a clear understanding of:

- Factors that affect the machinability of titanium and its alloys
- Mechanics of chip formation
- Appropriate cutting tools that last longer
- Modes of tool wear
- Application and effectiveness of coolant.

In order to study the above items, first, the term machinability must be defined. Machinability is usually used to define how easy or difficult the material can be cut or in other word be machined. This term refers to the ease of removing a layer of workpiece materials, which is usually metal, by means of a cutting tool. Machinability is not a parameter that can be directly measured using measurement units in terms of grades or numbers. For this reason, the machinability must be indirectly quantified based on the other parameters that can be measured. These parameters include:

- Power required to perform machining
- Surface finish generated by machining
- Tool life and modes of tool wear.

Typically, materials with good machinability rating can be machined relatively easily with much less power consumption. Machining such materials yields good surface quality (surface finish), and the cutting tools do not suffer from rapid wear and thus offer longer tool life. Unlike the materials with good machinability rating, those with low machinability rating consume high cutting power, cause rapid tool wear, and yield low-quality surface finish. During machining titanium and its alloys, the rapid tool wear is primarily because of the high abrasion between the tool and workpiece, and the highly localized temperatures at the cutting edge [32, 33]. Titanium's thermophysical properties are the root causes of several machining challenges.

3.6.1 Poor Thermal Conductivity

Titanium has a very low coefficient of thermal expansion and very poor thermal conductivity in comparison with other structural metallic materials such as iron, nickel, and aluminum. Table 3.4 shows the thermal properties of titanium versus the above-mentioned metals. As can be seen, thermal conductivity of α-titanium, which has the highest conductivity among the titanium alloys in the table, is one-fourth of that of iron and one-tenth of that of aluminum.

During machining operations, heat is generated usually in excessive amounts in the vicinity of cutting zone. Most of this heat is generated when the workpiece material is plastically deformed ahead of the cutting tool (primary shear zone). Plastic deformation consumes energy and converts it to heat. Heat is also generated when energy is consumed to overcome the friction along the chip–tool interface (secondary shear zone). Finally, another portion of heat is generated as the newly

Table 3.4 Thermal properties of titanium versus some other metals [2]

Material	Thermal conductivity ($W\ m^{-1}\ K^{-1}$)	Coefficient of linear thermal expansion ($10^{-6}\ K^{-1}$)
Ti–6Al–4V	7	9.0
Ti–15–3	8	8.5
α-Titanium	20	8.4
Fe	80	11.8
Ni	90	13.4
Al	237	23.1

machined surface is rubbed against the wear land. The wear land (tertiary shear zone) is along the flank face of the cutting tool. The heat generated during machining operation is beneficial as it locally softens the workpiece and leads to easier machining process. However, excessive heat, if not properly dissipated, is undesirable as it damages the cutting tool and adversely affects the quality and integrity of workpiece surface.

During machining operations, the generated heat is dissipated from the cutting system in four ways. A big portion of heat is transferred to the cutting tool, chip, and workpiece, while a smaller portion, usually negligible, is transferred to the surrounding environment. Low thermal conductivity of titanium and its alloys delays the heat dissipation process and prevents the effective heat dissipation from the shearing zone. Hence, the heat is mainly concentrated in the cutting zone [7]. The results of many researches and studies have demonstrated that during a typical machining operation the cutting tool receives up to 50% of the heat; however, this value drastically raises to 80% when titanium or its alloys are machined [7, 34, 35]. That means massive amount of heat is transferred to the cutting tool, which is definitely detrimental to the tool life. Figure 3.7 compares the distribution of heat between the cutting tool and the chip when machining Ti–6Al–4V and Steel CK45.

Moreover, machining titanium and its alloys produces thinner chips because of the high shear angle in the absence of built-up edge. Thinner chip is equivalent to a smaller contact area between the cutting tool and chip; thus, lower amount of heat can be extracted from the cutting zone and transferred to the chip [7]. In addition, the small contact area leads to higher stresses on the cutting tool [36–39]. All of the aforementioned factors contribute to the accumulation of heat within the cutting zone, which in turn elevates the temperature and leads to excessive tool wear as well as large plastic deformation on the cutting edge. Rapid deterioration of the cutting tool not only jeopardizes the productivity due to frequent stops for tool replacement, but also adversely affects the workpiece surface integrity. The increase in temperature also promotes diffusion wear along the rake face and cutting edge.

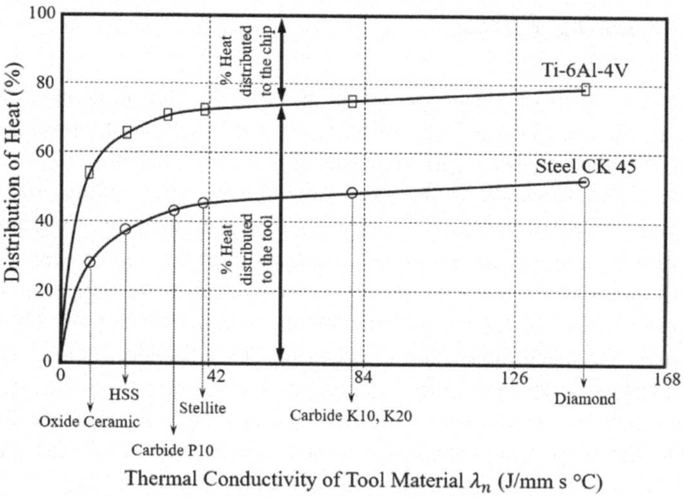

Fig. 3.7 Distribution of heat between the tool and the chip when machining Ti–6Al–4V and Steel CK45 (with permission to reuse) [7, 35]

3.6.2 Chemical Reactivity

Despite their chemical inertness at room temperature, titanium is highly reactive at high temperatures (>500 °C), which is a common temperature during machining; therefore, it can readily react with cutting tool materials [40, 41]. This undesirable feature is more dominant at high cutting speeds at which higher temperatures are possible. At elevated temperatures, titanium, as the workpiece material, can react with the coating layer of the cutting tool, which leads to a rapid removal of protection layer and leaves the substrate exposed and unprotected [7, 9]. Removal of coating makes the cutting tool very susceptible to chipping and premature tool failure [40].

3.6.3 Low Modulus of Elasticity

Titanium also has a low Young's modulus (modulus of elasticity) which leads to higher elastic deflection and spring back of the workpiece [42], especially in case of thin walls. This periodic deflection, while the tool proceeds over the workpiece surface, induces some vibrations and may result in chatter and consequently poor surface finish.

3.6.4 Hardening Effect

Hardening, as a result of plastic deformation and diffusion, is another factor that negatively affects the machinability of titanium and its alloys. Increasing the temperature of a system causes a corresponding increase in the diffusivity. During the cutting process, a large spike in the local temperature at the cutting zone causes a similar spike in diffusivity of workpiece material in the same vicinity. Cutting zone is normally surrounded by the air, which contains nitrogen and oxygen molecules. These molecules can diffuse into the titanium workpiece at temperatures in the range of 600–700 °C or higher and harden the surface layer [43] which consequently makes the machining more challenging. The diffusion may also occur at chip–tool interface where the chip is rapidly moving along the rake face. Due to increased temperature, the cobalt content of some cutting tools such as carbide tools diffuses into the chip. The cobalt acts as a binder for the tool material; hence, decreasing the cobalt content weakens the tool and shortens its life.

Furthermore, although the majority of metals fail to preserve their strength at elevated temperatures, titanium maintains its strength [44] at relatively high temperature, which further lowers its machinability.

3.7 Mechanics of Chip Formation

In previous sections, the titanium and its alloys are classified as difficult-to-cut material mostly because of their low thermal conductivity, hardening characteristics, low modulus of elasticity, and also their tendency to chemical reactions at high temperatures. In addition to all aforementioned factors, one more important feature must also be discussed when studying the machinability of titanium alloys. This important feature is the mechanics of chip formation, which directly affects the tool life and surface integrity during a typical machining process. To study the mechanics of chip formation, basic types of chips are briefly reviewed. There are several classifications for the different types of chips formed during machining operation; however, the chips can be divided into two main categories [45]:

- Continuous chips (steady state)
- Periodic or cyclic chips.

Periodic or cyclic chip can be further divided into four subcategories [45]:

- Discontinuous chips
- Wavy chips
- Built-up edge chips
- Saw-toothed chips.

Continuous chip is usually observed when shearing is concentrated along a thin shear plane and no plastic deformation or material flow occurred before the shear zone is reached or after the shear zone is passed. The formation of cyclic chips is more complex in comparison with the continuous ones. Among cyclic chips, discontinuous chips are formed because of crack initiation at the tool tip followed by crack propagation toward the free surface, while wavy chips are formed because of the cyclic change in shear angle. Built-up edge chips are formed when the workpiece material is gradually accumulated over the tip of the cutting edge until it becomes unstable and leaves the cutting edge.

The chip observed in titanium machining even in relatively low cutting speeds is classified as segmented chip with saw-toothed exterior. Studying the mechanism of chip segmentation is of high importance as it highly affects the machining parameters such as cutting forces and temperatures as well as surface integrity of the workpiece. Figure 3.8 compares the morphology of a saw-toothed chip versus a continuous one that is formed when machining Ti-6Al-4 V and steel, respectively.

Formation of the segmented saw-toothed chips can be explained by two theories:

- Adiabatic shear, i.e., thermoplastic shear instability
- Initiation and propagation of crack within the primary shear zone.

The first theory explains the segmented chip formation due to thermoplastic shear instability. The plastic instability is the result of simultaneous interaction between work hardening and thermal softening in the primary shear zone [46, 47]. One of the pioneering research works to explain the importance of thermoplastic shear instability on the mechanics of cyclic chip formation was presented by Recht [48]. The work demonstrated that the shear instability during machining occurs when the strengthening effect induced by the strain hardening is offset by the gradients of local temperature. The mechanics of chip segmentation during machining was further studied by Sowerby and Chandrasekaran who presented a model for the formation of saw-toothed chip in steels [49]. They related the formation of segmented chips to the accumulation of damage in the workpiece material during machining.

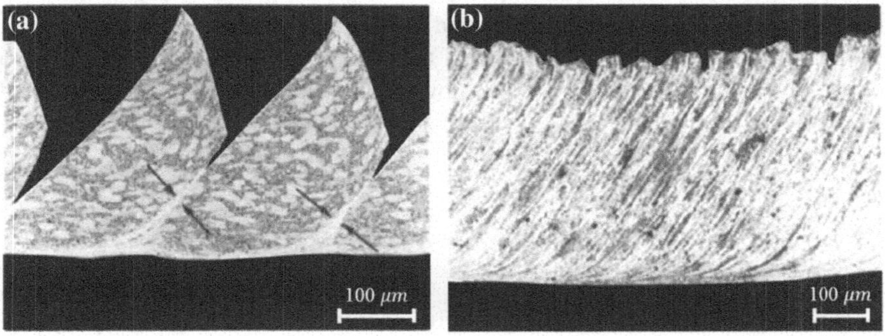

Fig. 3.8 Chip morphology when machining titanium and steel, **a** titanium Ti–6Al–4V saw-toothed chip and **b** steel continuous chip (with permission to reuse) [46]

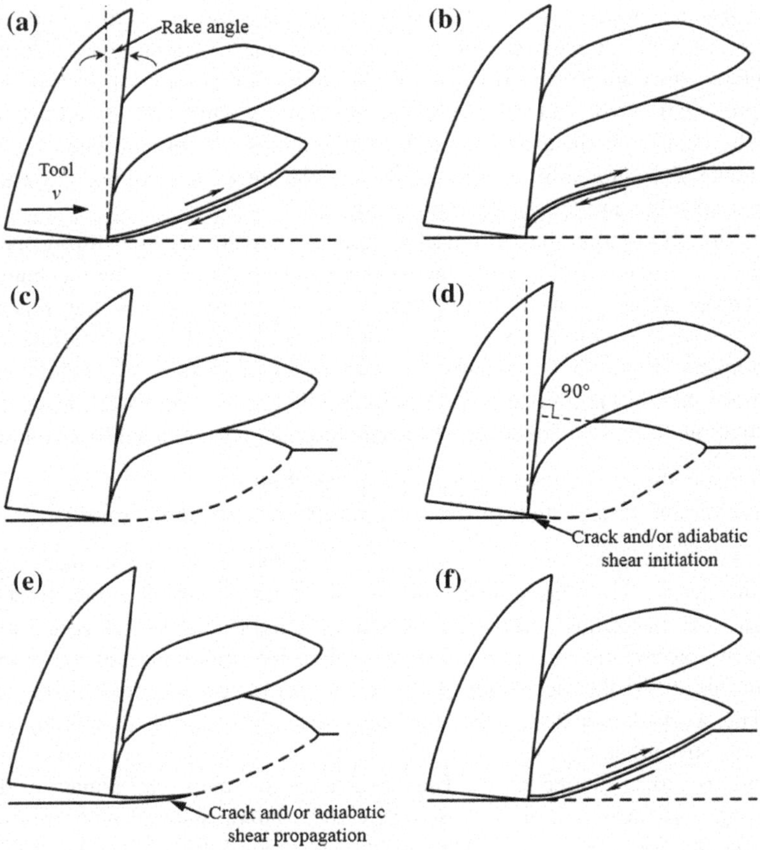

Fig. 3.9 Order of events during chip formation when machining titanium alloys (with permission to reuse) [46]

The second theory considers the crack initiation and crack propagation in the primary shear zone as the influential factors on the formation of segmented chip [45, 50]. Regardless of the theory behind the formation of segmented chips, this type of chip introduces cyclic variation of the machining forces, which may lead to tool chipping or breakage. Figure 3.9 shows the mechanics of chip formation and the chronological order of events when machining titanium alloys.

3.7.1 Chip Segmentation Under Adiabatic Shear

As mentioned earlier, during machining of titanium, the heat is not properly dissipated because of the poor thermal conductivity of the workpiece material. Thus, the heat is localized in the cutting zone and can be observed when examining the generated chips. Moreover, titanium and its alloys retain their hardness at relatively high temperatures that further diminish the positive effects of thermal softening in promoting their machinability. In such cases, the localized temperature only softens the material in the very close vicinity of the cutting zone and causes shear instability. The shear instability then acts as the driving force to minimize the work done along the shear plane. It has been shown that, during machining, the shear strength of workpiece material highly depends on strain (ε) and temperature (θ) and is governed by the following equation [48]:

$$\frac{d\tau}{d\varepsilon} = \frac{\partial\tau}{\partial\varepsilon} + \frac{\partial\tau}{\partial\theta}\frac{d\theta}{d\varepsilon} \tag{3.1}$$

The favorable condition for shear instability to occur is when the slope of the true stress/strain curve becomes zero. This condition can be estimated equating the above equation to zero. Thus, the catastrophic slip criteria in the primary shear zone can be presented in terms of:

$$0 \le \frac{\frac{\partial\tau}{\partial\varepsilon}}{-\frac{\partial\tau}{\partial\theta}\frac{d\theta}{d\varepsilon}} \le 1 \tag{3.2}$$

Based on Eq. (3.2), the catastrophic slip (shear) occurs if the above ratio has a value between 0 and 1. If the ratio becomes one, the catastrophic slip (shear) will be imminent [48]. When machining titanium and its alloys, especially at high speed, i.e., very high strain rates, a great amount of heat is generated along a thin cutting zone. Due to the low heat conductivity and the high rate of deformation, there is no enough time for the heat to dissipate away from the shear zone. The heat localization leads to thermal softening and consequently the catastrophic failure of the workpiece material. At the onset of catastrophic failure, the magnitude of critical shear strain is a function of temperature and strain rate where increasing the strain rate reduces the critical strain. As can be seen in Fig. 3.10, when machining titanium alloys (in this case Ti–6Al–4V commonly known as grade 5 titanium), continuous chip is formed when the strain rate is below the critical value and the segmented one is observed when the critical value is surpassed. Needless to mention here that high-speed machining is a relative concept depending on several factors such as the type of machining operation, the workpiece material, and also tool material, hence, a high cutting speed for a material with poor machinability may be considered as moderate or even low for another one with a better machinability.

As can be seen in Fig. 3.10, at strain rates higher than the critical value, segmented chips are formed instead of the continuous ones. This is mainly due to the fact that beyond the critical value, the effect of strain hardening on increasing the strength of material (strengthening) is seized by the effect of high cutting temperature on decreasing the strength (thermal softening). As a result, plastic deformation becomes unstable and leads to the formation of segmented chips. Investigating the

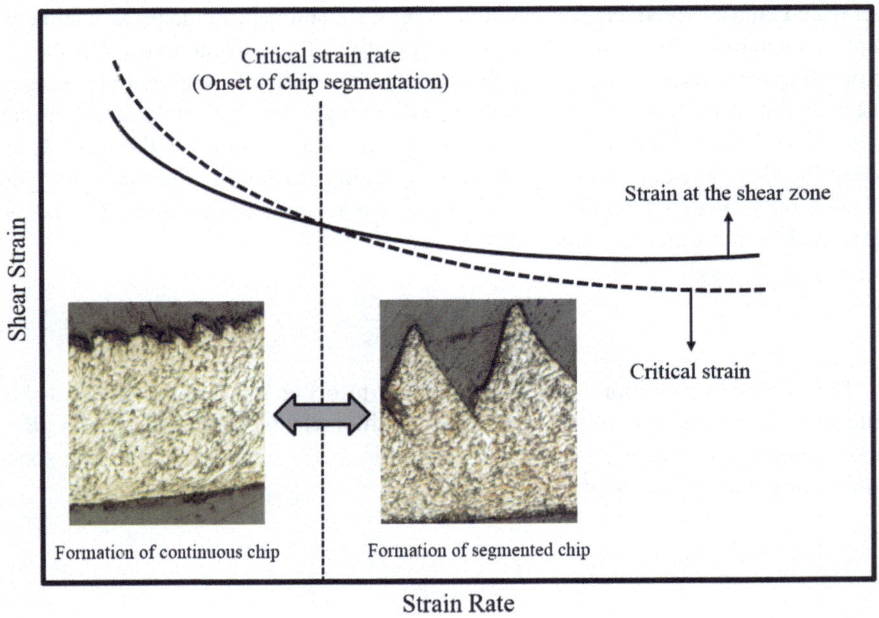

Fig. 3.10 Effect of the strain rate on the chip morphology when machining Ti–6Al–4V

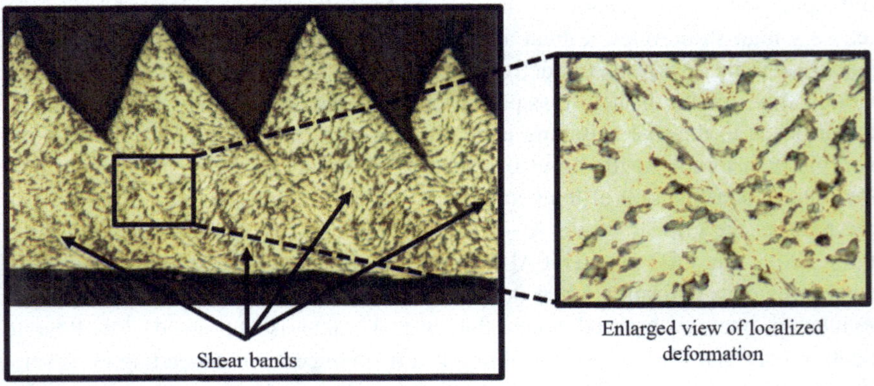

Fig. 3.11 Typical segmented chip demonstrating shear bands when turning Ti–6Al–4V

morphology of segmented saw-toothed chips in machining titanium alloys reveals localized deformation bands known as adiabatic shear bands. An enlarged view of bands of intensive shear that divide the chip into several segments is presented in Fig. 3.11.

As presented earlier, strain rate plays an important role in the formation of segmented chips when machining titanium and its alloys; hence, accurate calculation of strain rate is of prime importance. Strain rate when machining titanium is mainly calculated based on the definition presented by Komanduri [51]. Based on Komanduri's definition, heterogeneous deformation occurred within the chips while extreme shear is localized in the shear bands, which is very narrow. The concentration of intense shear in this thin area (shear band) splits the chip into detached sections [52]. The total shear strain (ε_t) can be calculated by considering both homogeneous shear (ε_h) in the chip segment and catastrophic shear (ε_c) in the shear band [52].

$$\varepsilon_t = \varepsilon_h + \varepsilon_c \tag{3.3}$$

In order to quantify the shear strain rate, Fig. 3.12 is considered. It illustrates an optical micrograph of a midsection of chips when machining Ti-5553.

Based on Fig. 3.12, homogeneous and catastrophic shear can be calculated as follows:

$$\varepsilon_h = \frac{\text{ae}}{\text{ad}} \tag{3.4}$$

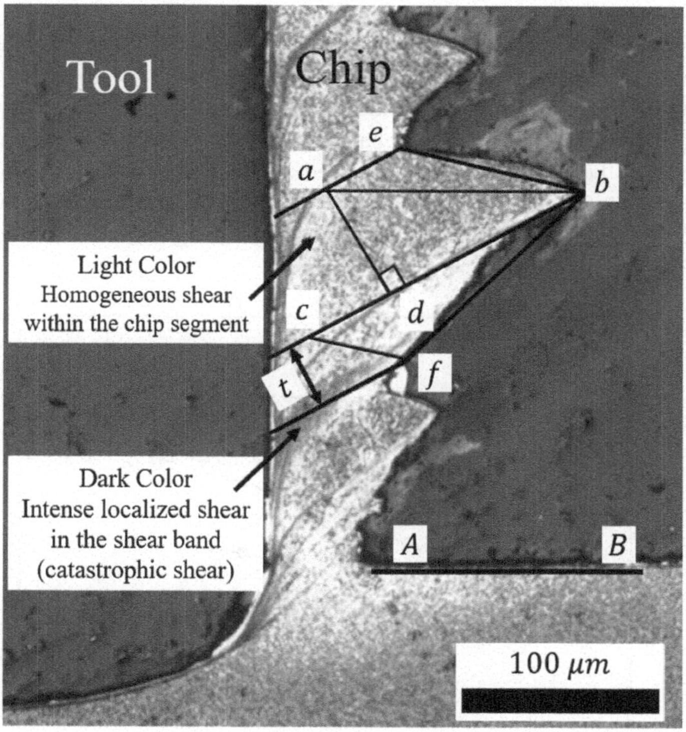

Fig. 3.12 Optical micrograph of a midsection of chips when machining Ti-5553 (with permission to reuse) [52]

$$\varepsilon_c = \frac{\text{cb}}{t} \tag{3.5}$$

In Eqs. (3.4) and (3.5), (t) is the shear band thickness and (cb) represents the shear displacement in the shear band. In Fig. 3.12, line (ab) is assumed to be parallel with line (AB). In the same fashion, (ae) is assumed to be parallel to (db) and (eb) is parallel to (cf) [52]. According to shear strain definition, before catastrophic shear band is formed, the two line segments (bf) and (eb) were parts of (ef). This is a realistic assumption as both sections have been exposed to the homogeneous shear before the formation of catastrophic shear. Therefore, it can be concluded that during chip formation, point (c) on the shear band travels to (b) and yields (cb) as a trace point which is equivalent to shear displacement [52]. The magnitude of this displacement is a measure to quantify the catastrophic shear. The same discussion was presented by Turley et al. [53] who presented the shear displacement in the shear band by line (cb). The total time to cross each segment of the chip including the homogeneous and catastrophic shear bands is equal to their overall thickness (ad + t) divided by cutting speed (v) [52]. As a result, strain rate can be calculated as follows:

$$\dot{\varepsilon}_t = \frac{\varepsilon_t}{(\text{ad} + t)/v} \tag{3.6}$$

Other important factors that affect the morphology of segmented chips are cutting speed and feed. The results of experimental works revealed that increasing cutting speed increases the segmentation distance; see Fig. 3.13. This is also demonstrated by less number of shear bands within the same length of the chip. The same trend has been observed for the feed. Figures 3.14 and 3.15 show SEM images of the generated chips when turning Ti–6Al–4V at different feeds.

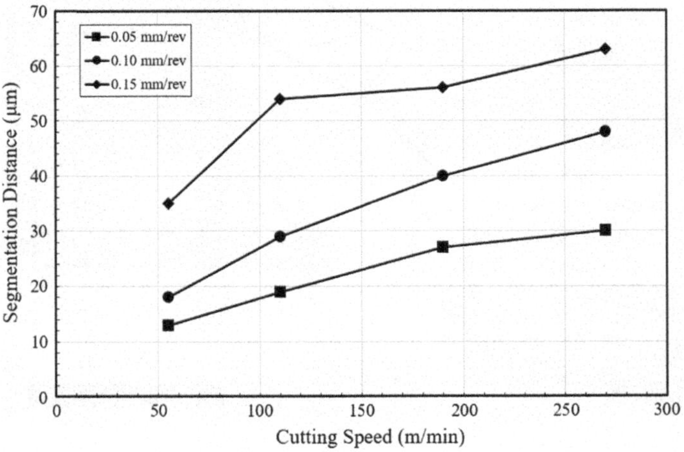

Fig. 3.13 Effect of cutting speed on the segmentation distance when turning Ti–6Al–4V

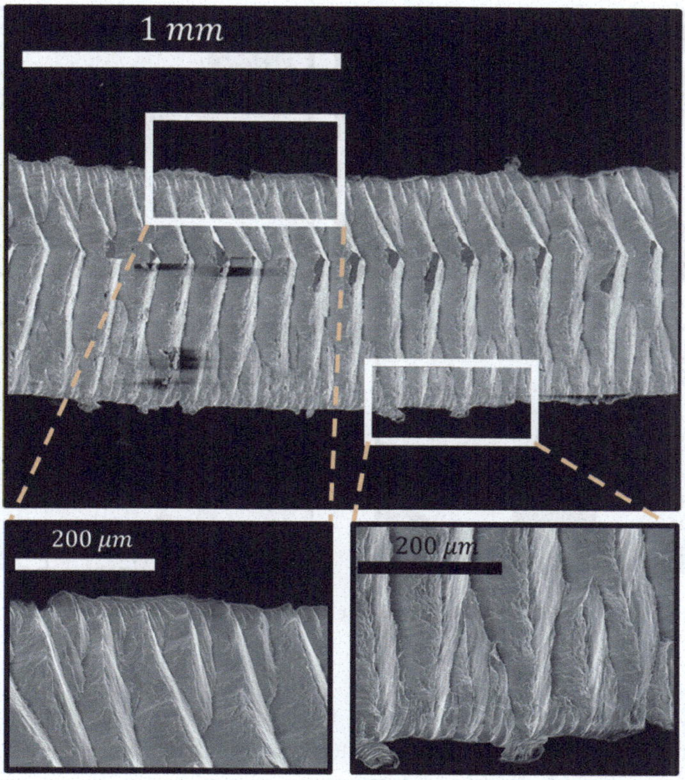

Fig. 3.14 SEM image of the free surface of a chip when turning Ti–6Al–4V at a high feed of 0.8 mm/rev

As clearly shown in Figs. 3.14 and 3.15, lower feed value yields a chip with narrower segments while higher feed values result in wider segments. In addition, when higher feeds are employed, a noticeable side flow of material occurs at the bottom of the chip close to the point where the tool nose separates the chip from the bulk of material. The plastic side flow, mainly perpendicular to the direction of chip flow velocity, is mainly caused by the nose radius that applies a high magnitude of pressure on a thin section of the workpiece material.

3.8 Appropriate Tool Materials and Modes of Tool Wear

To achieve an efficient process when machining titanium or any of its alloys, the cutting tool must demonstrate the following properties [41, 54]:

- Capability of retaining hardness at high cutting temperatures (hot hardness)
- Demonstration of good chipping resistance

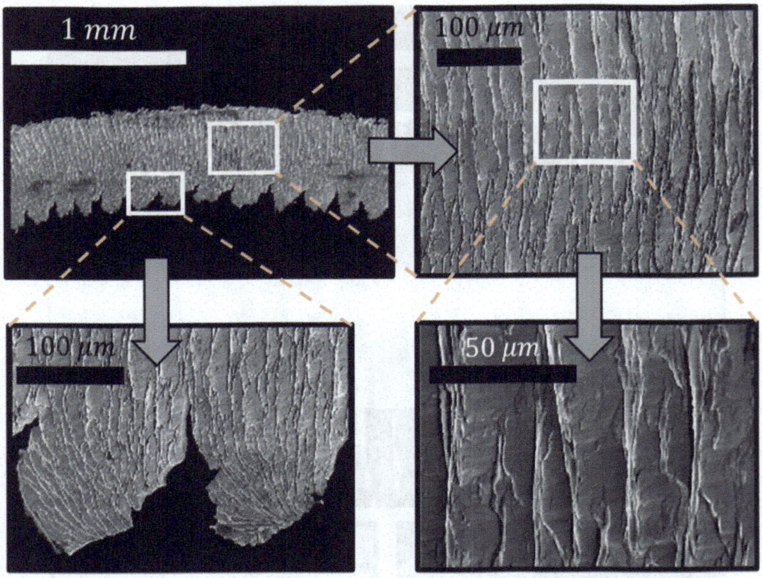

Fig. 3.15 SEM image of the free surface of a chip when turning Ti–6Al–4V at a low feed of 0.05 mm/rev

- Ability to withstand the cyclic cutting forces (toughness and fatigue resistance)
- Low chemical affinity with titanium (to prevent possible reactions)
- High thermal conductivity and compressive strength.

In this section, the feasibility of different cutting tool materials in machining titanium and its alloys is discussed.

3.8.1 HSS Tools

High-speed steel (HSS) tools are one of the oldest yet broadly used cutting tool materials in industry. They owe their reputation to their relatively high toughness, which makes them a primary candidate for interrupted or intermittent machining. However, these tools cannot be used in machining applications where the cutting temperature may reach or go beyond 500 °C [55]. This threshold is chosen due to the fact that the hardness of HSS tools rapidly drops in temperatures above 600 °C [56, 57]; therefore, the HSS cutting tools should not be exposed to temperatures even close to 600 °C.

Plastic deformation due to the rapid loss of hardness is considered as the dominant mode of tool wear when machining titanium and its alloys using HSS tools [7, 41]. This mode of tool wear is primarily originated from the presence of elevated temperatures accompanied by high compressive stresses. If not failed by plastic deformation, HSS tools greatly suffer from crater wear because of the high temperature generated along the chip–tool interface due to the relative motion between the chip and the tool rake face [41, 58]. The combination of these two wear modes makes the HSS tools deteriorate rapidly and shortens their life. Despite these facts, some grades of HSS tools such as highly alloyed, e.g., T5, T15, M33, and M40 series as well as general-purpose grades, e.g., M1, M2, M7, and M10 can be utilized in machining titanium [54]. However, the cutting speed must be kept below 30 m/min [54, 59–61], which is not a favorite speed in the era of high-speed machining and does not satisfy the high rate of productivity. Coating HSS tools can be a solution to make them practical for machining titanium alloys; nevertheless, not all coatings are suitable candidates since several coatings such as titanium nitride (TiN) and titanium carbonitride (TiCN) tend to react with titanium alloys during machining; thus, they must be avoided. Other coatings like chrome nitride (CrN) and titanium aluminum nitride (TiAlN) exhibit more favorable characteristics and can be utilized to protect HSS tools during machining titanium alloys. Generally speaking, although some grades of HSS tools are applicable in machining titanium alloys, this family of cutting tools is not the primary candidates for this application particularly with the current demand for high production rate.

3.8.2 Carbide Tools

Carbides (sintered/cemented) are a broad family of cutting tool materials made by binding hard carbide particles such as WC, TiC, TaCx, NbC, and Nb_2C in a cobalt matrix. [62, 63]. Carbide tools are available in different grades depending on the carbide particle contents. Among them, the straight grade cemented carbide (WC–Co) is preferred for machining titanium and its alloys. This grade includes 6 wt% cobalt (Co) and tungsten carbide (WC) grain size within the range of 0.8–1.4 μm [7, 36, 38, 54, 61, 64]. Carbide tools maintain their hardness at relatively higher temperatures than HSS tools, which allows cutting at higher speeds; however, their application in machining titanium and titanium alloys is limited to cutting speeds below 60 m/min [38]. Research investigations showed that when cutting speeds exceed this borderline, the carbide tools may experience plastic deformation as a result of heat generated in the cutting zone [36].

At cutting speeds lower than 45 m/min, the generated heat is not high enough to soften the carbide tools; as a result, plastic deformation and diffusion wear are not common in this cutting speed range. In such cases, the predominant modes of failure are microfractures and also mechanical and thermal fatigue. At higher cutting speeds, the temperature at the cutting zone may exceed 500 °C at which the titanium becomes highly reactive and diffusion wear starts. The process begins with the migration of titanium atoms from the workpiece and followed by the reaction of these atoms with carbon content of the cutting tool. The product of such a reaction is hard titanium carbide (TiC) layer that adheres to tool surface as well as workpiece [57, 61]. This layer may act as a barrier and prevents further diffusion. However, this layer also sticks to the workpiece and makes it harder to cut, which further encourages tool wear. Adhesion wear of carbide tools has also been observed when the generated cutting temperature increases beyond 740 °C. The interaction of these wear mechanisms results in combinations of wear modes such as crater and flank wear when titanium alloys are machined at high cutting speeds. Figure 3.16 shows the SEM images of a carbide tool tip used for machining of Ti-6Al-4V at relatively high cutting speed.

During metal machining, crater wear is initiated over the rake face particularly at a short distance from the cutting edge and it progresses as machining continues. The depth of crater wear gradually increases until the tool can no longer sustain the cutting conditions and fails. However, as can be seen in Fig. 3.16, when machining titanium and its alloys, the location of crater wear occurs almost on the tip of the cutting edge. Occurrence of crater wear over the cutting tip is attributed to the formation of segmented saw-toothed chips when machining titanium alloys. Formation of this type of chips significantly alters the nature of contact over the rake face (secondary shear/deformation zone) and pushes the crater wear toward the cutting edge. Crater wear on the tip of the cutting edge is drastically detrimental for

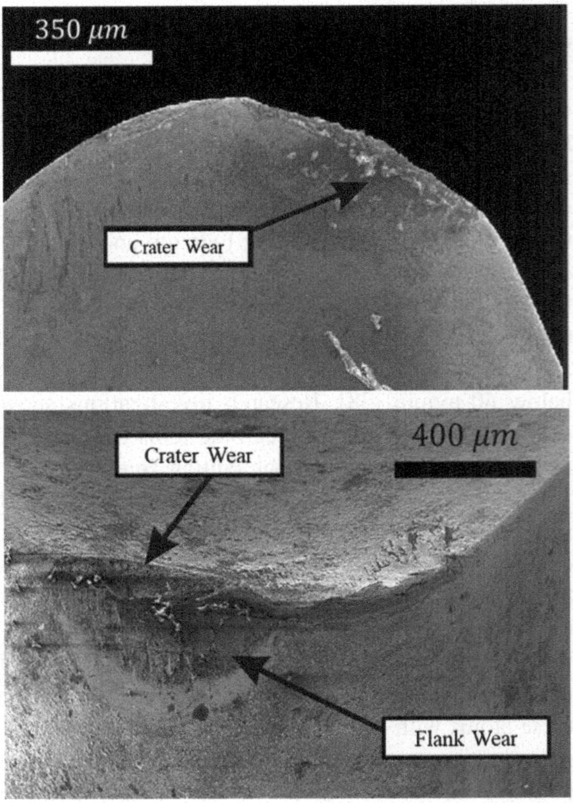

Fig. 3.16 SEM image of the tool tip showing crater wear (top) and both crater and flank wear (bottom) in high-speed turning of Ti–6Al–4V with carbide tool (TNMG 432-23)

both the cutting tool and machined surface quality. The cutting tool will be losing its distinctive geometry, especially in the case of form tools, and its strength while the surface quality deteriorates due to the effect of the damaged tool edge on the dimensional accuracy of the generated surface.

Coating of carbide tools adds an extra layer of protection and improves their performance when machining titanium and its alloys. Carbide tools protected by coatings show lower friction at the tool–chip interface; therefore, they experience lower cutting force and heat during cutting. However, the coated layer is quickly removed either by fast flow of chip over the rake face (abrasion) or by the chemical reaction between the workpiece (titanium) and the coating. Therefore, uncoated straight grade cemented carbides (WC–Co) are the most recommended carbide tools in machining operations when the titanium is involved [41]. More recent coating technologies offer a better adhesion to the carbide tools and thus increase the tool performance.

3.8.3 Ceramic Tools

Ceramic tools show hot hardness, higher compressive strength, chemical inertness, and better oxidation resistance [65] than HSS and carbide tools, which can be considered an advantage, while they are suffering from lack of toughness as well as mechanical and thermal shock resistance. Ceramic tools are very brittle and susceptible to breakage in interrupted or heavy cutting. When machining titanium alloys, ceramic tools drastically suffer from large groove wear and notch wear [38]. These limitations put a barrier for the application of ceramic tools in machining titanium and titanium alloys and make them not a recommended choice of tools for this application.

3.8.4 CBN and PCBN Tools

Stress, temperature, and vibration are the main reasons for tool damage, and they shorten the tool life during machining processes. These factors are more prominent when machining difficult-to-cut materials such as titanium and its alloys. As a result, if a cutting tool material can sustain high levels of stress, temperature, and vibration, it will be considered as the recommended option for machining titanium alloys.

CBN is the second hardest tool material next to the diamond. Its melting point is around 2730 °C, high enough to exhibit excellent hot hardness when exposed to high machining temperatures. CBN is also very stable up to 2000 °C with no sign of oxidation, while diamond which is harder starts to graphitize at temperatures around 900 °C [66]. Due to these superior features, cubic boron nitride (CBN) and polycrystalline cubic boron nitride (PCBN) tools are common candidates for machining difficult-to-cut materials; however, similar to ceramic tools when machining titanium, large groove wear is frequently observed on the flank face of CBN tools [67]. Hence, CBN tools are regarded as not appropriate tools for machining titanium and its alloys. However, polycrystalline cubic boron nitride (PCBN) and binderless cubic boron nitride (BCBN) tools are suitable candidates for machining titanium alloys [67, 68]. As expressed by their name, BCBN tools have no binder; hence, they demonstrate higher thermal conductivity and high-temperature durability [42, 59]. Application of CBN and PCBN tools in machining titanium and its alloy allows higher cutting speeds, much higher than those achievable by ceramics and carbide. Cutting speeds of 150 m/min for Ti–6Al–4V [69] and 185–220 m/min for $\alpha + \beta$ [70] have been reported using these types of tools.

As the hardness increases, toughness usually decreases; consequently, cutting tool materials of high hardness usually suffer from lack of toughness and shock resistance and CBN, PCBN, and BCBN tools are no exception. Nose wear, chipping, and low cycle fatigue due to cyclic mechanism of chip formation are among the common modes of tool wear when using the family of boron nitride tools when machining titanium. Diffusion in another mechanism of wear that can also be observed when a tool is exposed to nitrogen and oxygen from the surrounding atmosphere in the presence of high temperature. Generally, despite their great features and excellent capabilities, use of CBN and PCBN tools in machining titanium alloys is not as widespread as expected primarily due to their price that sometimes exceeds the price of their carbide counterparts by twenty times.

3.8.5 Diamond Tools

Scoring the first rank among all tool materials, diamond is the hardest cutting tool material available to the machining industry. In addition to its excellent hardness, diamond exhibits a low coefficient of friction accompanied by superior wear resistance. PCDs are made synthetically by bonding diamond particles (crystals) of different sizes using usually cobalt as a metallic bonder. The characteristics and thus the applications of PCD tools can be tailored by altering the grain size.

Due to their extreme hardness, PCD tools experience much lower wear rate than carbide tools when machining titanium alloys [71]. PCD tools have demonstrated satisfactory performance when machining compressor blades in aerospace industries, which is usually made from Ti–6Al–4V [72]. Other members of this family of cutting tools such as sintered diamond and natural diamond tools are among the candidates for machining titanium alloys. However, similar to the family of CBN tools, the major drawback toward widespread use of the diamond tools is their very high cost.

3.9 Application of Coolant in the Machining of Titanium

The low thermal conductivity of titanium and its alloys significantly affects the heat dissipation through the workpiece material and thus reduces machinability. The high temperature causes drastic deterioration of tool life and leads to poor machined surface quality. Therefore, the use of coolant to improve the heat dissipation and reduce the harmful effect of heat on cutting tools becomes essential in any titanium machining process. The coolants help improve surface finish and reduce residual stress. They also minimize the welding action that occurs between the cutting tool and the titanium at high temperatures and ultimately increasing the machining performance [40].

When using high-pressure coolant, it has been determined that tool life increases up to 300%. The high pressure allows for better penetration, through different areas in the cutting zone thus reduce the temperature of the cutting tool. As a result, the rate of diffusion wear is reduced; therefore, the life of the cutting tool is significantly extended. The coolant jet also reduces the welding of the chips to the tool due to the excessive heat and also washes the chips away from the cutting area [73].

Generally speaking, sufficient cutting fluid applied with high pressure and pointed toward the appropriate location (near the cutting tool tip) can be very effective to prolong tool life when machining titanium alloys. Acting as a coolant and a lubricant, the cutting fluid reduces the tool temperature, lowers the cutting forces, and prevents chip welding. Applying the right amount of cutting fluid at the cutting tool tip provides flushing action and removes chips from the cutting surface. It also minimizes thermal shock of milling tools and ignition of chips during grinding of titanium. High-pressure coolant supply can also result in small, discontinuous, and easily disposable chips.

In the case of high-pressure cooling, the coolant or in other word cutting fluid is brought to the cutting area by means of an external or through the tool nozzle, which is placed in a modified chip breaker on the tool. The coolant is transferred inside a vacuum sealed line and is released through a hole on the chip breaker onto the cutting edge at the contact area with the chip. The way the chip breaker is formed allows the chip to form away from the nozzle as to not impede the function of the nozzle and block the coolant flow [74].

The use of flood cooling raises both health and environmental concerns. This promotes other cooling techniques in which less amount of coolant is utilized. Liquid nitrogen or LN2 is a coolant with less environmental and health concern. With cutting temperatures reaching over 1000 °C, LN2 is an ideal coolant to be used. To remain liquid, the LN2 must remain below −195 °C, which would allow extreme cooling effect in the vicinity of the cutting zone. However, in order to keep the liquid nitrogen at its appropriate temperature, it must be stored in cryogenic liquid cylinders. These cylinders are insulated and vacuum sealed to keep a consistent pressure and have various relief valves and disks to keep the cylinder pressure below 350 psi. All of these factors are crucial when liquid nitrogen is being used as the cooling agent, but it also introduces another complication to the machining setup.

Another cryogenic coolants applicable to the machining of titanium are liquid hydrogen (must remain below −252 °C) and liquid oxygen (must remain below −182 °C). However, liquid nitrogen, which was discussed earlier, is the most frequently used one due to accessibility and availability. Unlike oxygen and hydrogen, liquid nitrogen is also safe to handle that makes it more convenient to use.

3.9.1 Utilization of Nano-cutting Fluids

Despite all the health and environmental concerns, the flood cooling is a typical cooling strategy used in many industries. Minimum quantity lubrication (MQL) has been implemented in different applications and proven to be successful. However, its cooling capacity is relatively lower than that of flood cooling. The addition of nano-additives is an effective way to increase MQL heat capacity. It improves the characteristics of the resultant nano-mist in terms of convection, conduction, and wettability. Generally, MQL nano-fluid offers two main advantages:

- It improves the thermal and friction behavior and thus improves the performance.
- It makes the process more sustainable by avoiding the use of flood cooling.

Figure 3.17 shows a schematic illustration of the MQL nano-fluid technique. As can be seen, the nano-fluid is atomized and thus forms a fine mist. The nanoparticles in the mist are surrounded by the base of the fluid film. These fine particles will penetrate into the chip–tool interface zone and forms a tribofilm, which helps absorbing the generated cutting heat. In addition, the employed nano-additives serve as spacers to reduce the friction along the tool and workpiece interface [75].

A research was conducted to study the influence of dispersed multi-wall carbon nanotubes (MWCNTs) on enhancing the MQL cooling and lubrication capabilities during the turning of Ti–6Al–4V alloy. Same machining tests were conducted

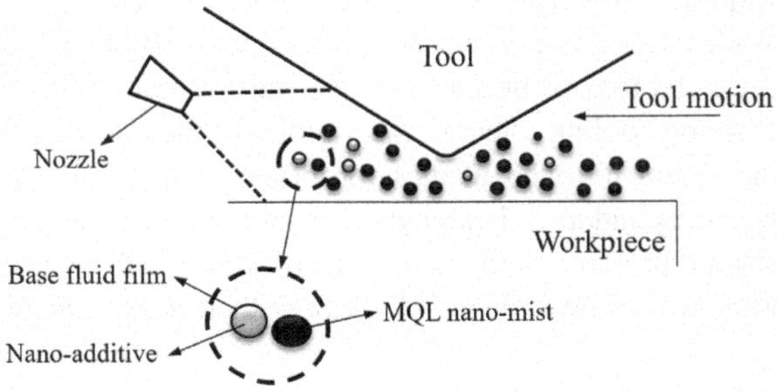

Fig. 3.17 MQL nano-cutting fluid mechanism

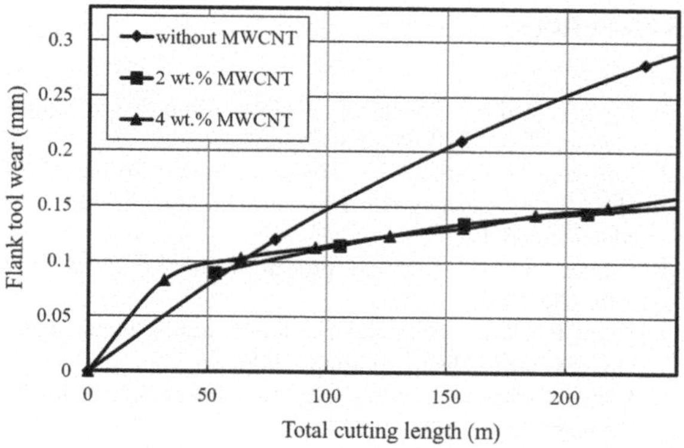

Fig. 3.18 Tool wear progression during machining of Ti–6Al–4V using different weight percentages of MWCNTs, $V = 120$ m/min, $f = 0.15$ mm/rev, and $d = 0.2$ mm

under different weight percentages of MWCNTs, and flank wear was measured after nine cutting passes (50 mm for each pass). The results revealed that the cutting using MWCNTs nano-fluids (2 and 4 wt%) has shown lower tool wear (see Fig. 3.18) compared to the tests conducted in the absence of nano-additives. Lower tool wear and consequently longer tool life can be attributed to the enhanced wettability, convection, and conduction of the fluid due to the presence of nano-additives. Improved cooling and lubrication properties of the resultant nano-fluid allow the cutting tool to retain its hardness for a longer time and thus increase the tool life.

3.10 Concluding Remarks

While titanium and its alloys are extremely versatile and useful for different industrial sectors, problems pertaining to their machining are a major roadblock toward their widespread applications. Despite their superior properties like corrosion and high-temperature resistance, lightweight and relatively long product life span, titanium alloys are still not widely used in industry, in comparison with aluminum and steels, mainly due to their high associated production costs. Although near net shape production is highly recommended because of its lower machinability, finish machining of titanium and its alloys is still inevitable in order to achieve tight tolerances and surface quality. With recent research into methods of improving production rate using hybrid machining techniques and creating new structural alloys with a focus on machinability, titanium still remains a unique material that cannot be easily utilized due to high cost and low machinability.

References

1. Encyclopaedia Britannica Online. http://www.britannica.com/EBchecked/topic/597135/titanium-Ti/ (2013). Accessed on May 2013.
2. Lütjering G, Williams JC. Titanium. Berlin: Springer; 2007.
3. Hosseini A, Kishawy HA, Hussein HM. Machinability of Titanium and Its Alloys. Mach Adv Mater. p. 95-118.
4. Gerdemann SJ. Titanium process technologies. Adv. Mater. Process. 2001;159 (DOE/ARC-2002-007).
5. Krebs RE. The history and use of our earth's chemical elements: a reference guide. Westport: Greenwood Publishing Group; 2006.
6. Warner J, Brandt D. Metallurgy fundamentals, vol. 69. Tinley Park: The Goodheart–Wilcox Company Inc.; 2005. p. 72–5.
7. Ezugwu E, Wang Z. Titanium alloys and their machinability—a review. J Mater Process Technol. 1997;68(3):262–74.
8. Joshi VA. Titanium alloys: an atlas of structures and fracture features. Boca Raton: CRC Press; 2006.
9. Yang X, Richard Liu C. Machining titanium and its alloys. Mach Sci Technol. 1999;3 (1):107–39.
10. Leyens C, Peters M. Titanium and titanium alloys: fundamentals and applications. New York: Wiley; 2003.
11. Xu L. Metastable beta titanium alloys: tuning the beta phase stability and low-temperature martensitic transformation of metastable beta titanium alloys. Delft: Delft University of Technology; 2015.
12. Welsch G, Boyer R, Collings E. Materials properties handbook: titanium alloys. Ohio: ASM International; 1993.
13. Poondla N, Srivatsan TS, Patnaik A, Petraroli M. A study of the microstructure and hardness of two titanium alloys: commercially pure and Ti–6al–4v. J Alloy Compd. 2009;486(1):162–7.
14. Weiss I, Semiatin S. Thermomechanical processing of alpha titanium alloys—an overview. Mater Sci Eng A. 1999;263(2):243–56.
15. Cited 2015; Available from: http://www.totalmateria.com/page.aspx?ID=CheckArticle&site=ktn&NM=218.
16. Donachie MJ. Titanium: a technical guide. Ohio: ASM International; 2000.
17. Chait R. Strength and toughness of a metastable beta-titanium alloy as influenced by strain rate variations. J. Test. Eval. 1976;4(5).
18. Oshida Y. Bioscience and bioengineering of titanium materials. New York: Elsevier; 2010.
19. Froes F, Suryanarayana C, Eliezer D. Synthesis, properties and applications of titanium aluminides. J Mater Sci. 1992;27(19):5113–40.
20. Cahn R. Intermetallic compounds as structural materials: history and prospects. Met Mater Process. 1989;1(1):1–19.
21. Anton D, Shah D, Duhl D, Giamei A. Selecting high-temperature structural intermetallic compounds: the engineering approach. JOM. 1989;41(9):12–7.
22. Callister WD, Rethwisch DG. Materials science and engineering: an introduction, 9th ed. New York: Wiley; 2013.
23. Szycher M. High performance biomaterials: a complete guide to medical and pharmaceutical applications. Boca Raton: CRC Press; 1991.
24. Schwartz M. Encyclopedia of materials, parts and finishes. Boca Raton: CRC Press; 2002.
25. Merlin PW. Design and development of the blackbird: challenges and lessons learned. AIAA. 2009;1522:2009.
26. NASA. http://www.nasa.gov/centers/dryden/history/pastprojects/SR71/index.html. Accessed on May 2013.

27. Inagaki I, Takechi T, Shirai Y, Ariyasu N. Application and features of titanium for the aerospace industry. Nippon Steel & Sumitomo Metal Technical Report. 2014;106:22–7.
28. Bewlay B, Nag S, Suzuki A, Weimer M. Tial alloys in commercial aircraft engines. Mater High Temp. 2016;33(4–5):549–59.
29. Schauerte O. Titanium in automotive production. Adv Eng Mater. 2003;5(6):411–8.
30. Davis JR. Metals handbook. Ohio: ASM International; 1998.
31. Elias C, Lima J, Valiev R, Meyers M. Biomedical applications of titanium and its alloys. JOM. 2008;60(3):46–9.
32. Li G, Sui X, Jiang C, Gao Y, Wang K, Wang Q, Liu D. Low adhesion effect of tao functional composite coating on the titanium cutting performance of coated cemented carbide insert. Mater Des. 2016;110:105–11.
33. Bai D, Sun J, Chen W, Du D. Molecular dynamics simulation of the diffusion behaviour between Co and Ti and its effect on the wear of Wc/Co tools when titanium alloy is machined. Ceram Int. 2016;42(15):17754–63.
34. Child H, Dalton A. ISI special report 94. London; 1968.
35. Konig W. Applied research on the machinability of titanium and its alloys. In: Proceedings of AGARD conference advanced fabrication processes, Florence, Italy; 1978.
36. Dearnley P, Grearson A. Evaluation of principal wear mechanisms of cemented carbides and ceramics used for machining titanium alloy ImI 318. Mater Sci Technol. 1986;2(1):47–58.
37. Motonishi S, Hara Y, Isoda S, Itoh H, Tsumori Y, Terada Y. Study on machining of titanium and its alloys. Kobelco Technol Rev. 1987;2:28–31.
38. Narutaki N, Murakoshi A, Motonishi S, Takeyama H. Study on machining of titanium alloys. CIRP Ann Manuf Technol. 1983;32(1):65–9.
39. Smart E, Trent E. Temperature distribution in tools used for cutting iron, titanium and nickel. Int J Prod Res. 1975;13(3):265–90.
40. Rahman M, Wong YS, Zareena AR. Machinability of titanium alloys. JSME Int J C. 2003;46:107–15.
41. Machado A, Wallbank J. Machining of titanium and its alloys—a review. Proc Inst Mech Eng B: J Eng Manuf. 1990;204(1):53–60.
42. Pramanik A. Problems and solutions in machining of titanium alloys. Int J Adv Manuf Technol. 2014;70(5–8):919–28.
43. He G, Zhang Y. Experimental investigations of the surface integrity of broached titanium alloy. CIRP Ann Manuf Technol. 1985;34(1):491–4.
44. Donachie M Jr. Introduction to Titanium and Titanium Alloys (Titanium and Titanium Alloys Source Book). Ohio: American Society for Metals; 1982. p. 7.
45. Vyas A, Shaw M. Mechanics of saw-tooth chip formation in metal cutting. J Manuf Sci Eng. 1999;121(2):163–72.
46. Komanduri R, Von Turkovich B. New observations on the mechanism of chip formation when machining titanium alloys. Wear. 1981;69(2):179–88.
47. Shaw M, Dirke S, Smith P, Cook N, Loewen E, Yang C. Machining Titanium. Cambridge: Dept. of Mechanical Engineering, Massachusetts Inst. of Tech.; 1954.
48. Recht R. Catastrophic thermoplastic shear. J Appl Mech. 1964;31(2):186–93.
49. Sowerby R, Chandrasekaran N. A proposal for the onset of chip segmentation in machining. Mater Sci Eng A. 1989;119:219–29.
50. Hua J, Shivpuri R. Prediction of chip morphology and segmentation during the machining of titanium alloys. J Mater Process Technol. 2004;150(1):124–33.
51. Komanduri R. Some clarifications on the mechanics of chip formation when machining titanium alloys. Wear. 1982;76(1):15–34.
52. Yan D, Hilditch T, Kishawy H, Littlefair G. On quantifying the strain rate during chip formation when machining aerospace alloy Ti-5553. Procedia CIRP. 2013;8:123–8.
53. Turley D, Doyle E, Ramalingam S. Calculation of shear strains in chip formation in titanium. Mater Sci Eng. 1982;55(1):45–8.
54. Siekmann HJ. How to machine titanium. Tool Eng. 1955;34(1):78–82.
55. Davim JP. Machining of hard materials. Berlin: Springer; 2011.

56. Kramer B. On tool materials for high speed machining. J Eng Ind. 1987;109(2):87–91.
57. Ezugwu E, Bonney J, Yamane Y. An overview of the machinability of aeroengine alloys. J Mater Process Technol. 2003;134(2):233–53.
58. Trent EM, Wright PK. Metal cutting. Oxford: Butterworth-Heinemann; 2000.
59. Rahman M, Wang Z-G, Wong Y-S. A review on high-speed machining of titanium alloys. JSME Int J C. 2006;49(1):11–20.
60. Pérez J, Llorente J, Sánchez J. Advanced cutting conditions for the milling of aeronautical alloys. J Mater Process Technol. 2000;100(1):1–11.
61. Hartung PD, Kramer B, Von Turkovich B. Tool wear in titanium machining. CIRP Ann Manuf Technol. 1982;31(1):75–80.
62. Upadhyaya GS. Cemented tungsten carbides: production, properties and testing. New York: William Andrew; 1998.
63. Davim JP. Machining: fundamentals and recent advances. Berlin: Springer; 2008.
64. Che-Haron C. Tool life and surface integrity in turning titanium alloy. J Mater Process Technol. 2001;118(1):231–7.
65. Chakraborty A, Ray K, Bhaduri S. Comparative wear behavior of ceramic and carbide tools during high speed machining of steel. Mater Manuf Processes. 2000;15(2):269–300.
66. Klocke F. Manufacturing processes. Berlin: Springer; 2011.
67. Pramanik A, Islam M, Basak A, Littlefair G. Machining and tool wear mechanisms during machining titanium alloys. In: Advanced Materials Research. Zürich: Trans Tech Publication; 2013.
68. da Silva RB, da Silva MB, Sales WF, Ezugwu EO, Machado ÁR. Advances in the turning of titanium alloys with carbide and superabrasive cutting tools. In: Advanced Materials Research. Zürich: Trans Tech Publication; 2016.
69. Ezugwu E, Da Silva R, Bonney J, Machado A. Evaluation of the performance of CBN tools when turning Ti–6al–4v alloy with high pressure coolant supplies. Int J Mach Tools Manuf. 2005;45(9):1009–14.
70. Zoya Z, Krishnamurthy R. The performance of CBN tools in the machining of titanium alloys. J Mater Process Technol. 2000;100(1):80–6.
71. Nabhani F. Machining of aerospace titanium alloys. Robot Comput Integr Manuf. 2001;17 (1):99–106.
72. Kuljanic E, Fioretti M, Beltrame L, Miani F. Milling titanium compressor blades with PCD cutter. CIRP Ann Manuf Technol. 1998;47(1):61–4.
73. Machado A, Wallbank J, Pashby I, Ezugwu E. Tool performance and chip control when machining Ti6al4v and Inconel 901 using high pressure coolant supply. Mach Sci Technol. 1998;2(1):1–12.
74. Hong SY, Markus I, Jeong W-C. New cooling approach and tool life improvement in cryogenic machining of titanium alloy Ti–6al–4v. Int J Mach Tools Manuf. 2001;41 (15):2245–60.
75. Hegab H, Umer U, Deiab I, Kishawy H. Performance evaluation of Ti–6al–4v machining using nano-cutting fluids under minimum quantity lubrication. Int J Adv Manuf Technol, 2018; p. 1–13.

Chapter 4
Superalloys

Abstract The term "superalloys" refers to a group of alloys that are capable of maintaining their mechanical characteristics after prolonged exposure to elevated temperatures. This category of material was primarily developed for applications such as turbo-superchargers and aircraft turbine engines. However, their applications have been expanded over the time to many other industrial sectors such as gas turbines, rocket engines, petroleum refineries, and chemical plants. From composition standpoint, among all of the metallic alloys ever developed for industrial, commercial, and military applications, superalloys are one of the most complex ones. This complexity enables metallurgists to develop different alloys and tailor their characteristics for wide range of applications. In addition to their modifiable features, the growing demand of industry for heat-resistant materials has further boosted superalloys development to the present level of sophistication. This chapter provides the readers with a brief review of superalloys, history of evolution, and their current industrial applications. It also presents the opportunities and challenges that may raise during machining superalloys. Applicable cutting tools, manufacturing processes, and other influential parameters on the machining and machinability of superalloys will also be discussed in this chapter.

4.1 Introduction

When safe performance at elevated temperatures is a design or functionality requirement, most of the materials and metallic alloys that are extensively used in industry lose their superiority and fail to satisfy the design standards. For instance, in power plants, which rely on superheated steam to produce electricity, some components are exposed to temperatures as high as high as 565 °C [1]. This range of temperature can be safely sustained by relatively cheap high-strength creep-resistant ferritic steels. However, these materials cease functioning when ultra-supercritical steam is used. Some components of such power plants may be exposed to 700 °C for a relatively long period of time which sometimes lasts for 200,000 h [1]. Being exposed to such severe service conditions, steels may experience higher corrosion attack. One may be tempted to consider other classes of

H. A. Kishawy and A. Hosseini, *Machining Difficult-to-Cut Materials*, Materials Forming, Machining and Tribology, https://doi.org/10.1007/978-3-319-95966-5_4

materials such as ceramics or titanium alloys for such applications. However, despite their very high hardness and great creep and oxidation resistance, ceramics such as silicon carbide and silicon nitride dramatically suffer from lack of toughness and ductility that make them very vulnerable to mechanical and thermal shocks. These limitations impose a barrier on their application in the above-mentioned working conditions. However, it is worth to mention that some types of ceramics such as those of zirconia-based are being used as thermal barrier coatings in high-temperature applications [1]. Titanium is another candidate for high-temperature applications, especially when lightweight is desired in applications such as aerospace as it offers great fatigue resistance and demonstrates high strength to weight ratio. However, its poor oxidation resistance limits its application in temperatures beyond 700 °C [1, 2]. Consequently, components made from titanium can only be used to produce lightweight parts to be utilized in applications in which parts are exposed to cooler temperatures like fans and compressors, not turbines where the temperature of inlet gases may reach 700 °C or even higher. When temperature is not a concern, the strength of materials is presented in terms of yield or ultimate tensile strength; nevertheless, these parameters are no longer the reliable indicators of strength when temperature is higher than 50% of the melting point/range [3]. At elevated temperatures, the strength becomes a function of time over which the material is subjected to a certain load. As a result, a material exposed to high temperature may start to elongate over the time until failure even under a load that is considerably lower than its room temperature strength. This time-dependent tendency of material to undergo permanent deformation is technically known as creep. High temperature also motivates and accelerates corrosion, which adversely affects the integrity of components over the time.

Without any notable counterpart, superalloys become the primary candidates when the strength and safe performance must be satisfied at elevated temperatures. Superalloys are undoubtedly "super" as they offer excellent high-temperature characteristics that enable engineers achieve reliability and cost-effectiveness. Superalloys are currently the most dependable and economically feasible means of achieving safe service life functionality at high operating temperatures in industrial gas turbines, components of aircraft engines, and heat exchangers. Although drawing a certain borderline for temperature above which utilization of superalloys becomes inevitable is not convenient, these materials are mostly utilized at temperatures beyond 540 °C [3]. Some others define superalloys as the materials well suited for applications at which the components may be exposed to 650 °C or higher temperatures for an extended period of time [4]. In general, superalloys are defined as a class of materials that have the ability to retain their mechanical characteristics even at homologous temperature $T_{operation}/T_{melting}$ (in °K) above 0.6 [1, 5]. They are considered as unique materials due to three desired characteristics [1, 5, 6]:

- Capability to operate at temperatures not far from their melting point
- Considerable resistance to mechanical degradation over prolonged period of time
- Resistance to environmental corrosion.

Superalloys are mainly categorized into three groups according to their primary ingredients, namely iron-based, nickel-based, and cobalt-based alloys. Among the three members of the superalloys family, iron-based alloys are the cheapest ones. However, despite their cost advantages, their applications are limited to lower-temperature applications [1, 6] due to inherent characteristics of iron. Cobalt-based superalloys exhibit superior characteristics at elevated temperatures; however, their widespread applications are hindered by nickel-based superalloys for two reasons. First, nickel is cheaper than cobalt; hence, it is more economically justifiable. The second reason is related to crystal structure [5] which will be further discussed later in this chapter.

4.2 Historical Background and Evolution of Superalloys

Design engineers and metallurgists are constantly challenged by the need of industry to design and develop high-strength materials with good corrosion and wear resistance for high-temperature applications. Although their existence is much older than the evolution of aircraft gas turbines, superalloys owe their growth and popularity to aircraft gas turbine engines. As a result, the history and evolution of superalloys are intertwined with the history and evolution of gas turbine engines.

Centuries ago, by observing the power of rising warm air, human recognized the direct relationship between efficiency and use of high temperatures. This observation eventually led to the evolution of thermodynamic principles such as the Brayton cycle, which is applied to rotating engines, stating that utilizing higher temperatures accompanied by lower heat loss yields more efficient operation [7]. However, it was not until late nineteenth and early twentieth century that relatively advanced steam turbines were introduced. Steam turbines evolution began in the 1800s, and gas turbines were introduced in the 1900s for power generation. The initial inventor of gas turbine engines is not clearly known. Credit, however, is mainly given to Dr. Frank Stolze, who filed a patent in 1872 describing a device close to what is currently known as gas turbine engine, and Charles Curtis, who produced a working model of such a device in the early 1900s [8]. Stolze's design and Curtis's model were both practically infeasible because the turbine produced less power than the compressor would require.

In 1903–1904, the first gas turbine engine was utilized in Paris, Europe, to generate electricity [7, 8]; while concurrently other concepts such as gasoline engines and flight were developing. This new technology, at that time, further progressed in 1905 through a joint collaboration between US Army, General Electric, and Cornell University. Directed and supervised by Dr. Stanford A. Moss that project led to the development of turbo-supercharger engines which were later employed in aircraft during World War I [7, 8]. This effort stimulated a further and continuous progress in the metallurgy of high-temperature alloys.

Later, Aegidius Elling, a Norwegian engineer, designed and built a gas turbine engine based on a centrifugal compressor and radial turbine. His gas turbine engine was capable of producing 11 hp at inlet temperature of approximately 400 °C (750 °F) [8]. This operational temperature was low enough so that austenitic stainless steels could be utilized to produce its structural components. Austenitic stainless steels had been previously discovered in the 1910s and were the primary candidate for high-temperature applications at that time [9]. Discovery of austenitic stainless steels was the result of research and observations done by H. Brearley in England and E. Maure in Germany during 1910–1915. Due to their relatively high-temperature resistance compared with other commercially available materials, gamma iron with face-centered cubic (FCC) structure became the favorite phase structure of choice upon which a new generation of high-temperature alloys, which were later named as superalloys, were developed [10, 11]. Although the term superalloy was not commonly used until World War II, the first known material that shows some characteristics of superalloys was an austenitic iron-based alloy, which contained 34% Ni, 11% Cr, and was hardened by 0.3% C.

In 1917, Imphy, a French company, filed a patent for an alloy that could be used in land-based gas turbines [12]. In addition to iron, this alloy comprised of 60% nickel, 11% chromium and was hardened by carbon and tungsten. In the same year, Pierre Chevenard, a French engineer and scientist at Imphy, filed another patent for an austenitic iron-based alloy specially developed to address the issue of intergranular corrosion in steam turbine blades. This alloy consisted of 34% nickel, 11% chromium and was hardened by 0.3% carbon [12].

In the 1920s and the early 1930s, Ni–Fe–Cr heat and corrosion-resistant alloys were produced using vacuum induction melting technology. Around 1918, a British patent introduced a nickel–chromium alloy named as Nichrome (Ni–20Cr) which can be considered as the origin for future superalloys such as Nimonics and Inconels [3, 8, 9]. Another major breakthrough occurred in 1929 when Bedford, Pilling, and Merica achieved a remarkable increase in creep resistance of nickel–chromium known as Nichrome (Nimonic 80) [7, 8]. In 1935, Elling presented an innovative gas turbine engine design capable of producing 75 hp without exceeding the turbine inlet temperature of 550 °C (1000 °F). This temperature was still within the acceptable range for safe operation of stainless steels; thus, there was still no immediate need for superalloys capable of maintaining their strength at elevated temperatures.

Until 1937, almost all of the gas turbine engines were mainly intended to generate electricity. Working for Ernest Heinkel, a German engineering company, Hans von Ohain and Max Hahn developed and produced the first gas turbine engine for aircraft in 1937. Independent from von Ohain and Hahn, Sir Frank Whittle, developed and patented Whittle's engine in 1939 [7, 8]. By that time, the peak operating temperature of gas turbine engines was no more than 705 °C. The introduction of Heinkel's engine and Whittle's engine was a turning point in the history of superalloys as the inlet temperature of gas turbines had increased to 780 °C. The intensive research and development in turbine engines have led the scientists and engineers to conclude that the efficiency of gas turbine engines could be theoretically increased by increasing the temperature. However, conventionally available austenitic stainless steels at that era were not capable of sustaining that high-temperature range, which made structural integrity a crucial limiting factor. The knowledge and understanding of the mechanical behavior of metallic alloys at the higher operating temperatures were also limited at that time. Figure 4.1 shows the historical development of superalloys up to World War II.

Between 1930s and 1950s, research studies on the development of superalloys mainly focused on microstructure optimization which was then followed by process optimization [8]. Driven by World War II, development of superalloys experienced a swift boost in 1940s. This major development was primarily driven by the early applications of jet engines in military aircraft and then by the need of industry to heavy duty yet strong and reliable gas turbines. The introduction of jet engines to aerospace industry was the result of a radical change in designers thinking which was initiated by advances in aerodynamic theory. These advancements made it apparent that [7]:

- Two-thirds of the input power to conventionally driven aircraft wasted due to turbulent drag.
- Supersonic forces at the end of propellers would prevent aircraft to reach speeds beyond 650 km per hour.

Many scientific studies were performed to develop high-temperature alloys during World War II, and these efforts built the basic concepts of new alloys [12]. During the era between 1940s and 1950s, extensive developments were made which resulted in the development of many new alloys particularly during World War II. The same trend continued in 1960s. The superalloys produced in those decades were usually used as cast due to ease in production. The first vacuum-melted alloys were introduced around 1955 following the invention of commercial vacuum melting technology by Darmara in 1952. Vacuum melting technology provided the metallurgists with the opportunity to keep the charge molten for a longer period of time which in turn helped to effectively degasify the

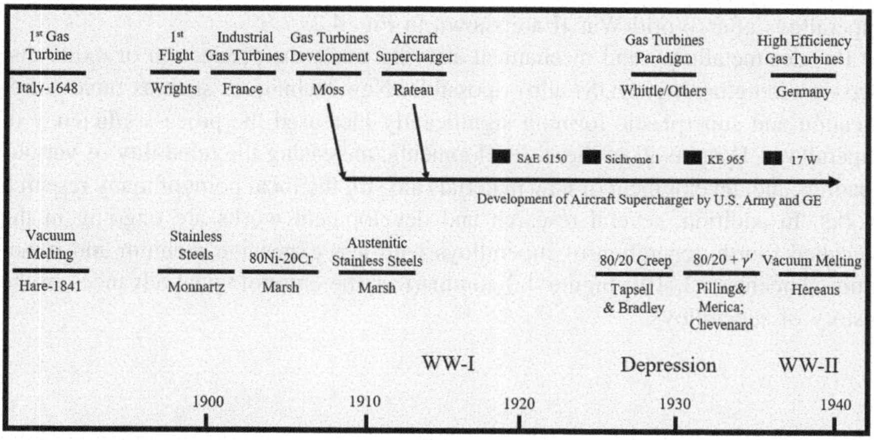

Fig. 4.1 Historical development and application of superalloys in their early stages [7]

melt and remove the impurities [8, 13, 14]. Superalloys were further developed in 1960s by the introduction of oxide dispersion strengthened (ODS) alloys at Dupont in 1965, mechanically alloyed alloys at INCO in 1966 and directionally solidified alloys for airfoils by Pratt and Whitney in 1969 [15].

In addition to higher levels of alloying elements, which were made possible by the introduction of vacuum casting processes, some issues such as detrimental phases were also introduced. These undesirable phases mainly formed because of high chromium content and caused the superalloys to lose some of their desired characteristics during the service life. This issue urged the metallurgists to reduce the chromium content from approximately 20 to 10 wt% which then presented another problem. It was observed that although lowering the chromium content reduced the chance of formation of the detrimental phases, it made the superalloys vulnerable to hot corrosion [8]. The issue was addressed by the application of coating to increase the environmental resistance of superalloys without jeopardizing their mechanical properties.

The history of superalloys will never be complete without mentioning the coatings. The first generation of coatings was developed in the 1950s and the 1960s to protect alloys from oxidation and corrosion at high temperature (hot corrosion). Successively, in the following decades, coatings were further developed not only for corrosion resistance but also for heat protection. Thermal barrier coatings (TBCs) were produced by plasma spray, and then by electron beam physical vapor [12]. The first generation of coatings was mainly composed of aluminum which was diffused in large amount into the surface of the superalloy material as the substrate [8, 16]. Application of aluminum coating was highly beneficial and increased the oxidation and corrosion resistance of superalloys used in turbine inlets up 1200 °C (2200 °F). Coating of the first-stage turbine vanes made of cobalt-based superalloys was done in the late 1950s. This can be considered the first commercial application of coating to protect superalloys from oxidation and corrosion. The developed technology was then further employed for coating turbine blades made of nickel-based superalloys in mid-1960s [17, 18]. Some of the major developments in superalloys after World War II are shown in Fig. 4.2.

Powder metallurgy and mechanical alloying made the production of oxide dispersion strengthening (ODS) alloys possible. New techniques such as rapid solidification and superplastic forming significantly increased the process efficiency of superalloys. Despite all of these developments, increasing the reliability of various coatings and development of new materials are still the focal point of many research works. In addition, several research and development works are ongoing in the so-called fourth generation of superalloys with less expensive rhenium and ruthenium elements [12, 19]. Figure 4.3 summarizes the chronological advances in the history of superalloys.

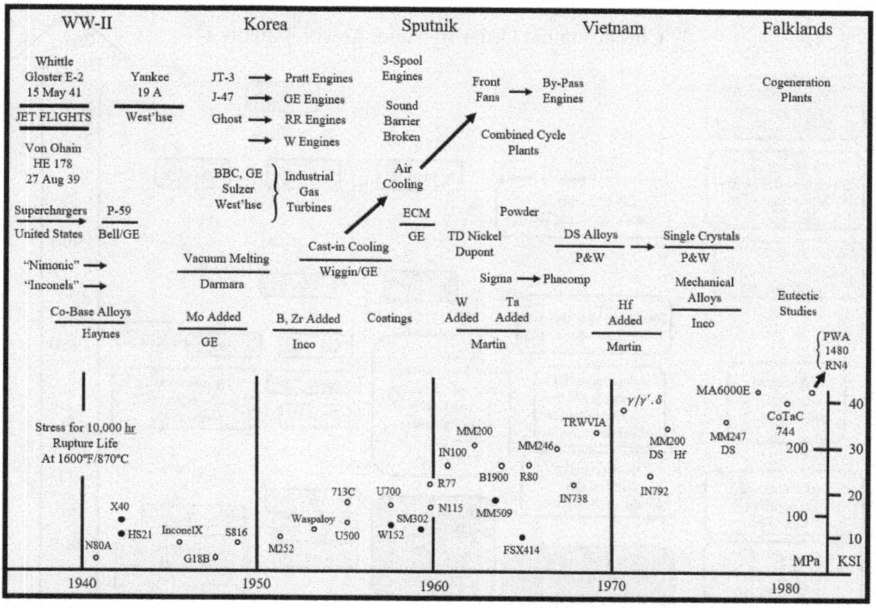

Fig. 4.2 Historical development of superalloys during and after World War II [7]

4.3 Metallurgy of Superalloys

As mentioned earlier, superalloys are mainly used for applications when enduring high temperatures and creep resistance are major design concerns. These alloys are based on Group VIII B elements of the Periodic Table [20], and they mainly contain Fe, Ni, Co, as well as additives such as chromium (Cr), tungsten (W), molybdenum (Mo), tantalum (Ta), niobium (Nb), titanium (Ti), and aluminum (Al). To optimize the metallurgical characteristic and achieve the favorable properties, the process of producing superalloys must be performed under controlled conditions. These controlled conditions yield desired mechanical properties of superalloys such as high-temperature, corrosion, and oxidization resistance along with high strength. As previously mentioned in the text, superalloys include three main classifications, namely iron–nickel-based superalloys, nickel-based superalloys, and cobalt-based superalloys [21]. These categories can be further subdivided into cast and wrought macrostructures [3]. Superalloys can also be studied at atomic level based on their crystal structure or their microstructure, which can be observed under microscope. In the following sections, metallurgy of different families of superalloys and present phases in their microstructure will be presented.

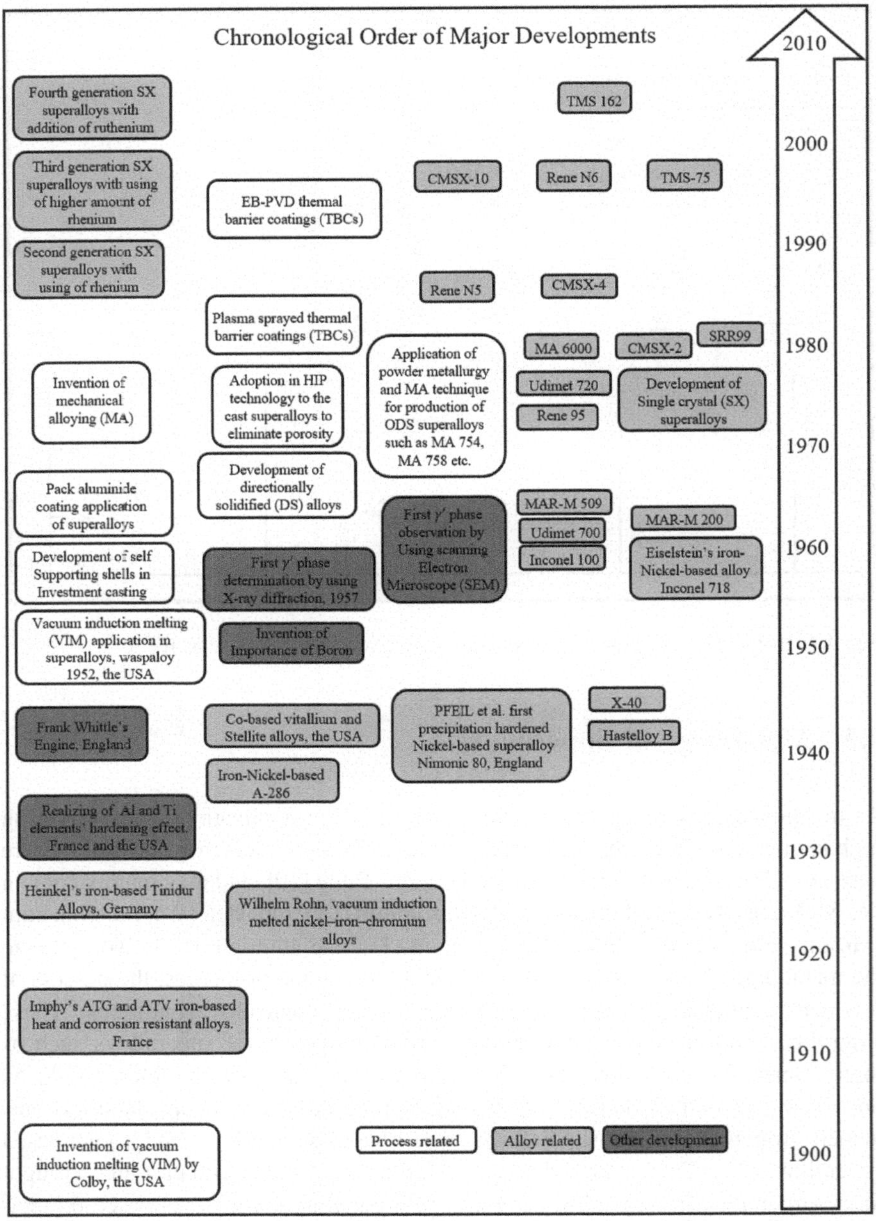

Fig. 4.3 Some of the major developments in superalloys [12]

4.3.1 Phases of Superalloys

The crystal structure of iron, nickel, and cobalt is generally face-centered cubic (FCC austenitic) which is the basis for superalloys. Superalloys primarily have austenitic face-centered cubic (FCC) matrix called gamma (γ) phase (this phase is suitable for solid solution and dispersion strengthening, and it is very ductile) [12] plus a variety of secondary phases. However, the normal room temperature structures of iron and cobalt elemental metals are not FCC. Iron has body-centered cubic (BCC) crystal structure, and cobalt has hexagonal close-packed (HCP) structure at room temperatures. Both iron and cobalt transform and become FCC at high temperatures or in the presence of other elements alloyed with iron and cobalt. Alternatively, nickel has FCC crystal structure at all temperatures. Thus, in the case of superalloys based on iron and cobalt, the FCC forms are generally stabilized by the addition of alloying elements, particularly nickel, to provide the best properties [22]. Secondary phases of importance in controlling properties are carbides, gamma prime (γ'), gamma double prime (γ''), eta (η), and delta (δ) [3].

4.3.1.1 Gamma (γ) Phase

Gamma (γ) phase with FCC structure is the primary phase of superalloys. It is a desirable phase due to its ductility and toughness. FCC structure has a lower diffusivity, due to its close-packed structure, compared to BCC structure. This structure is the key to superior characteristics and the stability of superalloys at high temperatures [5].

4.3.1.2 Gamma Prime (γ') Phase

Nickel and iron–nickel alloys owe their strength primarily to their intermetallic phase gamma prime (γ') which is one of the secondary phases present in superalloys microstructure. This phase is formed when the appropriate amount of aluminum and titanium are added to precipitate γ' (Ni_3Al, Ti) phase with FCC crystal structure [23]. Other elements such as niobium, tantalum, and chromium also enter the gamma prime phase [23]. The chemical compatibility along with close match in matrix/precipitate lattice parameter ($\sim$0–1%) enables the γ' phase to precipitate homogeneously all over the matrix and exhibits high stability even at elevated temperatures up to almost 1100 °C. Gamma prime phase is also ductile and coherent with matrix [5, 24, 25]; hence, it increases the matrix strength without jeopardizing its fracture toughness. This phase is the necessary factor for creep resistance plus high-temperature strength of superalloys.

4.3.1.3 Gamma Double Prime (γ'') Phase

Gamma double prime (γ'') is another secondary phase in superalloys microstructure which is usually found in nickel–iron superalloys. This phase includes a combination of nickel and niobium in the presence of iron. This combination forms a body-centered tetragonal (BCT) crystal structure named as Ni_3Nb which is coherent with the γ matrix [23]. This phase is not stable at temperatures beyond 650 °C (1200 °F); therefore, it only guarantees the high strength of superalloys at low to intermediate temperatures [23]. The strength of superalloys is mainly derived from precipitated phases and solid solution hardeners. Gamma prime (γ') and gamma double prime (γ''), which are usually found in iron–nickel-based and nickel-based superalloys, are the principal strengthening precipitate phases.

4.3.1.4 Carbides

Metal carbides (MC) are combinations of carbon in the amount of approximately 0.02–0.2 wt% and reactive and refractory elements like titanium, tantalum, hafnium, and niobium. At elevated temperatures, metal carbides have a tendency to precipitate from the liquid phase. Carbides also have a tendency to decompose during heat treatment and form other carbides like $M_{23}C_6$ or M_6C that are known as lower carbides. Lower carbides are more likely to form on the grain boundaries and they increase rupture strength at elevated temperatures [23]. Carbides are found in all three superalloy groups.

Although, as mentioned above, γ' and γ'' are the key precipitate phases directly responsible for strengthening superalloys; however, carbides may also provide partial strengthening effects directly or indirectly. It either directly increases the strength of superalloys by dispersion hardening or indirectly by stabilizing grain boundaries against excessive shear [3].

4.3.2 Strengthening Mechanisms

As mentioned above, superalloys are strengthened mainly by the basic nature of the FCC matrix and its chemistry; however, this is not the only factor affecting the strength of these unique materials. Superalloys can also be strengthened by the presence of special strengthening phases, usually precipitates. Mechanical deformation, usually cold working, can also increase the strength of superalloys, but that increase in strength may not endure at elevated temperatures [22]. Some tendency toward transformation of the FCC phase to stable lower-temperature phases occurs in cobalt-based superalloys. The austenitic FCC matrices of superalloys have significant solubility for some alloying additions, which is a favorable characteristic for the precipitation of uniquely effective strengthening phases.

Several strengthening mechanisms/strategies can be adopted to improve the yield strength and creep resistance (deformation at elevated temperatures) of superalloys. Strengthening in superalloys can be achieved through solid solution hardening (substituted atoms) and work hardening [26]. Another strengthening mechanism is precipitation hardening because precipitates usually restrict the deformation. Strength can also be enhanced by distribution of secondary phases such as carbides in the entire metal matrix production to interfere with deformation. This strengthening mechanism is particularly effective in cobalt-based superalloys as it blocks dislocations and slip along the grain boundaries.

The most effective element for solid solution hardening is rhenium (Re); however, molybdenum (Mo), tantalum (Ta), and tungsten (W) also offer hardening effects. Chromium (Cr) and cobalt (Co) have moderate hardening effects; however, chromium increases oxidation resistance and cobalt lowers stacking fault energy [27]. There are also limits to the amount of solid solution hardening elements that can be added to the alloy, as they are heavy (not desirable for aerospace applications) and can form unwanted types of intermetallic compounds with each other if the concentration is above a certain threshold.

Another method of strengthening superalloys is precipitation hardening in which the alloy is solution annealed, quenched, and reheated to an intermediary temperature for a certain amount of time. This produces the desired mechanical properties in the superalloy such as high strength and corrosion, oxidization and creep resistance at high temperatures.

The greatest strengthening effect comes from the precipitates of γ' in the FCC γ matrix [3]. This is a classic FCC crystal structure for an alloy. They are cubic due to the specific low interface energy on the face of the crystal structure; however, the lattice parameters are not equal. This causes a strain at the interface, which is responsible for the dislocation barrier. An interesting consequence about the Ni_3Al precipitate is that it gets stronger as the temperature increases [28]. This is interesting, as most metals that undergo a temperature increase will have increasing bond length in their crystal structures that allow dislocations to pass through them and let them undergo plastic deformation at lower elongation.

Another strengthening mechanism in superalloys is known as superdislocation pairs. These pairs include two dislocations that are closely spaced and have a combined movement. It is a very high-energy defect arrangement and contributes to the strength of the Ni_3Al precipitate [28]. Table 4.1 summarizes the principal strengthening mechanisms of superalloys.

Table 4.1 Principal strengthening mechanisms of superalloys [12]

Solid solution strengthening	Its effectiveness is directly related to the atomic size difference between solute and solvent. This method introduces stability at elevated temperatures. It is also very effective in improving creep and stress rupture resistance. However, it is not very efficient on mechanical strength. Elements such as Mo, Co, and W are important solid solution hardener elements
Precipitation hardening of carbides	It is mainly used for strengthening cobalt-based superalloys. As a result, the carbon content of cobalt-based superalloys is relatively higher than that of other types of superalloys. Higher carbon content results in brittleness, and it is considered as a disadvantage of carbide strengthening
Precipitation hardening of intermetallics	It is the key strengthening mechanism of nickel- and iron-based superalloys
Grain size	Grain size is an adjustable parameter, and hence, it is very important. The success of strengthening mechanism by controlling grain size is highly depends on the temperature and thickness of part. Large grains lead to less grain boundary in the superalloys and are typically desirable in gas turbine parts
Grain orientation	When applied loads are anisotropic (e.g., turbine blades), controlling the grain orientation is a very useful strengthening mechanism. Grains can be oriented in desired direction using directional solidification for cast alloys and rolling for wrought alloys
Grain aspect ratio (GAR)	Grain aspect ratio is the ratio of grains' long axis to short axis. This ratio is particularly important in mechanically alloyed oxide dispersion strengthening series of superalloys
Crystallographic orientation	Some orientations such as <100> results in higher strength in FCC matrix and thus they are preferable. Such an orientation can be obtained by single crystal casting
Alloy cleanliness	The properties of superalloys can be significantly altered by cleanliness. It usually reduces incipient melting temperature and formation of unwanted phases

4.4 Detailed Classification of Superalloys

Superalloys are divided into three categories based on the dominant metal present in the alloy. These three categories are

- Iron-based superalloys
- Nickel-based superalloys
- Cobalt-based superalloys.

4.4.1 Iron-Based Superalloys

Iron-based superalloys are chronologically the first category of superalloys that were initially developed from austenitic stainless steels. Iron-based superalloys are cheaper than cobalt or nickel-based superalloys. They can be divided, in terms of their strengthening mechanism, into three groups. These groups include iron-based alloys strengthened by martensitic transformation, austenitic alloys strengthened by a series of metalworking either cold or hot, and finally austenitic alloys strengthened by precipitation hardening [29]. However, the first two categories are usually considered, by some metallurgists, as high-temperature, high-strength alloys and only the latter is recognized as a superalloy. Among the first two categories, the martensitic alloys are usually implemented at relatively low temperatures (lower than 1000 °F), while the austenitic types are used above 1000 °F [29]. The iron-based superalloys are usually suitable for use up to temperatures no more than 760 °C (1400 °F) [21].

One of the important characteristics of stainless steels, which iron-based superalloys evolved from, is their corrosion and oxidation resistance, which can be achieved by adding more than 10 wt% of chromium. This family of superalloys is formed as a combination of a close-packed FCC matrix with both solid solution hardening and precipitate-forming elements [29]. The first iron-based superalloy was made by adding titanium to high-chromium-content austenitic stainless steels to obtain age hardening [12]. To stabilize their FCC austenitic matrix, iron-based superalloys are commonly alloyed with minimum 25% nickel. They are usually composed of nickel ranging from 25 to 60% and iron ranging from 15 to 60% [12]. The alloys in this category with considerable nickel content are sometimes called nickel–iron-based superalloys [3].

Additional alloying elements like molybdenum (Mo) and chromium (Cr) are also added to this family to achieve solid solution hardening. In addition to its solid solution hardening effect, chromium effectively increases the oxidation and sulfidation resistance of this group of alloys within their service temperature range [12]. These desired characteristics are achieved by formation of a very thin and adherent chromium-rich oxide layer, which acts as a barrier to protect stainless steels in corrosive environments. In order to reach this level of protection, the Cr content must be high enough (higher than 9 wt%) to effectively cover the alloy surface with continuous oxide layer [12]. Solid solution hardening of this family of superalloys can also be improved by the addition of alloying elements with small atomic radius such as carbon (C), nitrogen (N), and boron (B).

Since iron is an affordable element, vast attempts have been made in academia and industry to further develop this category of superalloys. However, it must be noted that although iron is relatively cheaper than nickel and cobalt and may lead to substantial cost saving and improved forgeability, the presence of iron adversely affects the corrosion resistance [12]. Table 4.2 summarizes the AISI 600 series of superalloys consists of six subclasses of iron-based alloys, and Table 4.3 demonstrates some common types of iron-based superalloys.

Table 4.2 AISI 600 series of superalloys consists of six subclasses of iron-based alloys [29]

	Type	Code
Martensitic	Martensitic low alloy steels	601–604
	Martensitic secondary hardening steels	610–613
	Martensitic chromium steels	614–619
	Semi-austenitic and martensitic precipitation-hardening stainless steels	630–635
Austenitic	Austenitic steels strengthened by hot/cold work	650–653
	Austenitic superalloys; all grades except alloy 661 are strengthened by second-phase precipitation	660–665

Table 4.3 Examples of iron-based superalloys. *Source of data* [21]

Iron-based superalloys	
Alloy	Chemical composition
A-28	26% Ni, 15% Cr, 1.3% Mo, 0.2% Al, 2% Ti, 54% Fe, 1.3% Mn, 0.5% Si, 0.05% C, 0.015 B
Discaloy	26% Ni, 13.5% Cr, 2.7% Mo, 0.1% Al, 1.7% Ti, 54% Fe, 0.9% Mn, 0.8% Si, 0.04% C, 0.005% B
Alloy 901	42.5% Ni, 12.5% Cr, 5.7% Mo, 0.2% Al, 2.8% Ti, 36% Fe, 0.1% Mn, 0.1% Si, 0.05% C, 0.015% B
Haynes® 556[a]	20% Ni, 22% Cr, 20% Co, 3% Mo, 2.5% W, 0.1% Nb, 0.3% Al, 29% Fe, 1.5% Mn, 0.4% Si, 0.1% C
Incoloy® 800	32.5% Ni, 21% Cr, 0.4% Al, 0.4% Ti, 46% Fe, 0.8% Mn, 0.5% Si, 0.05% C
Incoloy® 801	32% Ni, 20.5% Cr, 1.1% Ti, 44% Fe, 0.8% Mn, 0.5% Si, 0.05% C
Incoloy® 802	32.5% Ni, 21.5% Cr, 8% Co, 46% Fe, 0.8% Mn, 0.4% Si, 0.4% C
Incoloy® 807	40% Ni, 20.5% Cr, 0.1% Mo, 5% W, 0.2% Al, 0.3% Ti, 25% Fe, 0.5% Mn, 0.4% Si, 0.05% C
Incoloy® 825[b]	42% Ni, 22% Cr, 15% Co, 3% Mo, 0.2% Al, 0.8% Ti, 28% Fe, 1% Mn, 0.5% Si, 0.05% C
Incoloy® 903	38% Ni, 13% Co, 3% Nb, 0.7% Al, 1.4% Ti, 41% Fe
Incoloy® 907	38% Ni, 13% Co, 4.7% Nb, 0.03% Al, 1.5% Ti, 42% Fe, 0.15% Si
Incoloy® 909	38% Ni, 13% Co, 4.7% Nb, 1.5% Ti, 42% Fe, 0.4% Si, 0.01% C, 0.001% B
N-155[C]	20% Ni, 21% Cr, 20% Co, 3% Mo, 2.5% W, 1% Nb, 30% Fe, 1.5% Mn, 0.5% Si, 0.15% C
V-57	27% Ni, 14.8% Cr, 1.3% Mo, 0.3% Al, 3% Ti, 52% Fe, 0.3% Mn, 0.7% Si, 0.08% C, 0.01% B
19-9 DL	9% Ni, 19% Cr, 0.4% Co, 1.3% W, 0.3% Ti, Fe (bal), 1% Mn, 0.5% Si, 0.3% C
16-25-6	25.5% Ni, 16.25% Cr, 6% Mo, Fe (bal), 2% Mn, 1% Si, 0.1% C
Pyromet® CTX-1	37.7% Ni, 0.1% Cr, 16% Co, 0.1% Mo, 3% Nb, 1% Al, 1.7% Ti, 39% Fe, 0.03% C
Pyromet® CTX-3	38.3% Ni, 0.2% Cr, 13.6% Co, 4.9% Nb, 0.1% Al, 1.6% Ti, Fe (bal), 0.15% Si, 0.05% C, 0.007% B
17-14CuMo[d]	14% Ni, 16.0% Cr, 2.5% Mo, 0.4% Nb, 0.3% Ti, 62.4% Fe, 0.75% Mn, 0.5% Si, 0.12% C
20-Cb3[e]	34% Ni, 24% Cr, 2.5% Mo, 1% Nb, 35% Fe, 0.07% C

[a]2% N, 0.02% La, and 0.9% Ta
[b]2.2% Cu
[c]15% N
[d]3% Cu
[e]3.3 Cu

4.4.2 Nickel-Based Superalloys

Nickel-based alloys are the most prominent class of superalloys in terms of strength and temperature resistance. Nickel is a tough metal with great corrosion and oxidization resistance in air and seawater [30] which is often used to make alloys with iron, molybdenum, and copper. Nickel is obtained from nickel-rich ores and is refined to produce pure nickel and nickel oxide as well as an alloy of iron and nickel called ferronickel. These nickel products can be mixed with other metallic elements to create nickel alloys. Specialized types of nickel alloys are made to resist corrosion, high stresses, and very high temperatures (as hot as 1050 °C). The temperature may even exceed this level and locally reaches 1200 °C at hot spots near airfoil tips [31]. This temperature is almost 90% of the nickel's melting point, which is around 1455 °C. The ability of nickel to sustain such severe conditions has made them a primary candidate for producing components of high-performance gas turbines. Nickel-based superalloys are made of a variety of elements, namely nickel, chromium, cobalt, iron, niobium, tungsten, and molybdenum [21]. Most of nickel-based superalloys rely on the nickel–chromium–aluminum phase diagram for their strengthening properties. A wide variety of microstructures exists in different nickel-based superalloys. In a single-phase gamma matrix, there exist gamma prime and carbide precipitates. The microstructure of the superalloy depends on its composition, service history, and thermal and mechanical treatment.

Some common types of nickel-based superalloys are Inconel 600 and Inconel 718 [32]. Inconel 600 is alloy strengthened to produce a tensile strength of 95 ksi, yield strength of 45 ksi, Young's modulus of 30 ksi, and melting point of 1355 °C to 1413 °C. It has excellent strength and exceptional oxidation resistance at high temperature, but with lower corrosion resistance than austenitic stainless steels [21]. Inconel 718 is a precipitate hardened superalloy with tensile strength of 180 ksi, yield strength of 150 ksi, Young's modulus of 30.6 ksi, and melting point of 1260–1336 °C [32]. It is one of the strongest superalloys at low temperature and the most commonly used one among superalloys. However, Inconel 718 suffers from lack of strength at elevated temperature and fails to maintain its strength in the range of 650–815 °C [21]. Inconel X-750 is another example of nickel-based superalloys, which is the precipitate-hardening version of Inconel 600. The yield strength of Inconel X-750 is three times higher than that of Inconel 650 at a temperature of 540 °C. A small number of nickel-based superalloys are produced with other forms of strengthening, such as solution hardening and oxide dispersion hardening.

Nickel-based superalloys exhibit many unique mechanical properties, such as low-temperature toughness, high-temperature strength, and exceptional corrosion strength in many different environments. Nickel-based superalloys are highly tough at lower temperatures due to their FCC crystal structure. The thermal diffusivity of nickel-based superalloys is relatively low, which makes it more likely to distort when welding because the heat is retained instead of dissipating quickly through the base metal. The thermal expansion coefficients of nickel-based superalloys are about the same as those for carbon and low alloy steels [21]. Table 4.4 shows different nickel-based superalloys.

Table 4.4 Examples of nickel-based superalloys. *Source of data* [21]

Nickel-based superalloys	
Alloy	Chemical composition
Astroloy	55% Ni, 15% Cr, 17% Co, 5.3% Mo, 4% Al, 3.5% Ti, 0.06% C, 0.03% B
Cobalt 214	75% Ni, 16% Cr, 4.5% Al, 2.5% Fe, 0.01% Y
D-979	45% Ni, 15% Cr, 4% Mo, 1% Al, 3% Ti, 27% Fe, 0.3% Mn, 0.2% Si, 0.05% C, 0.01% B
Hastelloy® C-22	51% Ni, 22% Cr, 2.5% Co, 13.5% Mo, 4% W, 5.5% Fe, 1% Mn, 0.1% Si, 0.01% C, 0.3% V
Hastelloy® C-276	55% Ni, 15% Cr, 2.5% Co, 16% Mo, 3.7% W, 5.5% Fe, 1% Mn, 0.1% Si, 0.01% C, 0.3% V
Hastelloy® G-30	43% Ni, 29.5% Cr, 2% Co, 5.5% Mo, 2.5% W, 0.8% Nb, 15% Fe, 1% Mn, 1% Si, 0.03% C, 1.7% Cu
Hastelloy® S	67% Ni, 15.5% Cr, 14.5% Mo, 0.3% Al, 1% Fe, 0.5% Mn, 0.4% Si, 0.009% B, 0.05% La
Hastelloy® X	47% Ni, 22% Cr, 1.5% Co, 9% Mo, 0.6% W, 18.5% Fe, 0.5% Mn, 0.5% Si, 0.1% C
Inconel® 587	Ni (bal), 28.5% Cr, 20% Co, 0.7% Nb, 1.2% Al, 2.3% Ti, 0.05% C, 0.003% B, 0.05% Zr
Inconel® 597	Ni (bal), 24.5% Cr, 20% Co, 1.5% Mo, 1% Nb, 1.5% Al, 3% Ti, 0.05% C, 0.012% B, 0.05% Zr, 0.02% Mg
Inconel® 600	76% Ni, 15.5% Cr, 8% Fe, 0.5% Mn, 0.2% Si, 0.08% C
Inconel® 601	60.5% Ni, 23% Cr, 1.4% Al, 14.1% Fe, 0.5% Mn, 0.2% Si, 0.05% C
Inconel® 617	54% Ni, 22% Cr, 12.5% Co, 9% Mo, 1% Al, 0.3% Ti, 0.07% C
Inconel® 625	61% Ni, 21.5% Cr, 9% Mo, 3.5% Nb, 0.2% Al, 0.2% Ti, 2.5% Fe, 0.2% Mn, 0.2% Si, 0.05% C
Inconel® 706	41.5% Ni, 16% Cr, 2.9% Nb, 0.2% Al, 1.8% Ti, 40% Fe, 0.2% Mn, 0.2% Si, 0.03% C
Inconel® 718	52.5% Ni, 19% Cr, 3% Mo, 5.1% Nb, 0.5% Al, 0.9% Ti, 18.5% Fe, 0.2% Mn, 0.2% Si, 0.03% C
Inconel® X-750	73% Ni, 15.5% Cr, 1% Nb, 0.7% Al, 2.5% Ti, 7% Fe, 0.5% Mn, 0.2% Si, 0.04% C
M-252	55% Ni, 20% Cr, 10% Co, 10% Mo, 1% Al, 2.6% Ti, 0.5% Mn, 0.5% Si, 0.15% C, 0.005% B
Nimonic® 75	76% Ni, 19.5% Cr, 0.4% Ti, 3% Fe, 0.3% Mn, 0.3% Si, 0.1% C
Nimonic® 80A	76% Ni, 19.5% Cr, 1.4% Al, 2.4% Ti, 0.3% Mn, 0.3% Si, 0.06% C, 0.003% B, 0.06% Zr
Nimonic® 90	59% Ni, 19.5% Cr, 16.5% Co, 1.5% Al, 2.5% Ti, 0.3% Mn, 0.3% Si, 0.07% C, 0.003% B, 0.06% Zr
Nimonic® 105	53% Ni, 15% Cr, 20% Co, 5% Mo, 4.7% Al, 1.2% Ti, 0.3% Mn, 0.3% Si, 0.13% C, 0.005% B, 0.1% Zr
Nimonic® 115	60% Ni, 14.3% Cr, 13.2% Co, 4.9% Al, 3.7% Ti, 0.15% C, 0.16% B, 0.04% Zr
Nimonic® 263	51% Ni, 20% Cr, 20% Co, 5.9% Mo, 0.5% Al, 2.1% Ti, 0.4% Mn, 0.3% Si, 0.06% C, 0.001% B, 0.02% Zr

(continued)

Table 4.4 (continued)

Nickel-based superalloys	
Alloy	Chemical composition
Nimonic® 942	39.5% Ni, 12.5% Cr, 6% Mo, 0.6% Al, 3.7% Ti, 37% Fe, 0.2% Mn, 0.3% Si, 0.03% C, 0.01% B
Nimonic® PE.11	37.5% Ni, 18% Cr, 5.2% Mo, 0.8% Al, 2.3% Ti, 35% Fe, 0.2% Mn, 0.3% Si, 0.05% C, 0.03% B, 0.2% Zr
Nimonic® PE.16	43% Ni, 16.5% Cr, 1% Co, 1.1% Mo, 1.2% Al, 1.2% Ti, 33% Fe, 0.1% Mn, 0.1% Si, 0.05% C, 0.02% B
Nimonic® PK.33	56.0% Ni, 18.5% Cr, 14% Co, 7% Mo, 2% Al, 2% Ti, 0.3% Fe, 0.1% Mn, 0.1% Si, 0.05% C, 0.03% B
Pyromet® 860	43% Ni, 12.6% Cr, 4% Co, 6% Mo, 1.25% Al, 3% Ti, 30% Fe, 0.05% Mn, 0.05% Si, 0.05% C, 0.01% B
Rene® 41	55% Ni, 19% Cr, 11% Co, 1% Mo, 1.5% Al, 3.1% Ti, 0.09% C, 0.005% B
Rene® 95	61% Ni, 14% Cr, 8% Co, 3.5% Mo, 3.5% W, 3.5% Nb, 3.5% Al, 2.5% Ti, 0.15% C, 0.01% B, 0.05% Zr
Udimet® 400	60% Ni, 17.5% Cr, 14% Co, 4% Mo, 0.5% Nb, 1.5% Al, 2.5% Ti, 0.06% C, 0.008% B, 0.06% Zr
Udimet® 500	54% Ni, 18% Cr, 18.5% Co, 4% Mo, 2.9% Al, 2.9% Ti, 0.08% C, 0.006% B, 0.05% Zr
Udimet® 520	57% Ni, 19% Cr, 12% Co, 6% Mo, 1% W, 2% Al, 3% Ti, 0.05% C, 0.005% B
Udimet® 630	60% Ni, 18% Cr, 3% Mo, 3% W, 6.5% Nb, 0.5% Al, 1% Ti, 18% Fe, 0.03% C
Udimet® 700	55% Ni, 15% Cr, 17% Co, 5% Mo, 4% Al, 3.5% Ti, 0.06% C, 0.03% B
Udimet® 710	55% Ni, 18% Cr, 15% Co, 3% Mo, 1.5% W, 2.5% Al, 5% Ti, 0.07% C, 0.02% B
Udimet® 720	55% Ni, 17.9% Cr, 14.7% Co, 3% Mo, 1.3% W, 2.5% Al, 5% Ti, 0.03% C, 0.033% B, 0.03% Zr
Unitemp AF2-1DA6	60% Ni, 12% Cr, 10% Co, 2.7% Mo, 6.5% W, 4% Al, 2.8% Ti, 0.04% C, 0.015% B, 0.1% Zr, 1.5% Ta
Waspaloy	58% Ni, 19.5% Cr, 13.5% Co, 4.3% Mo, 1.3% Al, 3% Ti, 0.08% C, 0.006% B

4.4.3 Cobalt-Based Superalloys

The third group of superalloys is cobalt-based superalloys. Cobalt has FCC crystal structure at room temperature which transforms to HCP at temperatures above 417 °C [30]. The major difference between cobalt-based superalloys and their iron-based and nickel-based counterparts is the absence of ordered coherent precipitates γ' or γ'' [12]. Cobalt-based superalloys are extensively strengthened by combining carbides and solid solution hardeners [12, 21]. The main alloying elements with cobalt are iron, nickel, and chromium; hence, common types of carbide in cobalt-based superalloys are M_6C, M_7C_3, and $M_{23}C_6$. MC-type carbide cannot be formed in cobalt-based superalloys as they do not contain tantalum (Ta), titanium (Ti), zirconium (Zr), or hafnium (Hf) [12].

Cobalt-based superalloys have substantial wear, heat, and corrosion resistance, but they exhibit lower strength and creep resistance compared to nickel-based superalloys but they are capable of maintaining their strength up to relatively higher temperatures [29]. Thus, their applications are similar to, but less rigorous than those of nickel-based ones [21]. Similar to nickel-based superalloys, cobalt-based superalloys are mainly used in applications within the temperature range of 650–1150 °C (1200–2100 °F), yet their applications are not as common as nickel-based superalloys. They are more expensive than iron-based and nickel-based superalloys, which imposes an extra obstacle to their wide application.

There are many types of cobalt-based superalloys. The most common ones are Haynes 25, UMCo-50, and Stellite 6B. Haynes 25 is considered as the well-known cobalt-based wrought superalloy that is formed by machining, cold working, welding, and forging. It has outstanding temperature strength and galling resistance with oxidization resistance up to 980 °C. It is also resistant to marine environments, acids, and body fluids. UMCo-50, consisting of 21% iron, is another common cobalt-based superalloy. It is not as strong as Haynes 25; therefore, it has smaller scale applications. The third example of cobalt-based superalloys is the high-temperature Stellite 6B, which has strong resistance to oxidization mainly due to its chromium content. It has high hot hardness because of the formation of complex carbides. Stellites are mostly used for applications where wear resistance and high-temperature strength are important concerns. They are the most common superalloys used in conditions that require highly wear-resistant material at high velocity and elevated temperatures [21]. Table 4.5 shows different cobalt-based superalloys.

Table 4.5 Examples of cobalt-based superalloys. *Source of data* [21]

Cobalt-based superalloys	
Alloy	Chemical composition
Air resist 213	19% Cr, 66% Co, 4.7% W, 6.5% Ta, 3.5% Al, 0.18% C, 0.15% Zr, 0.1% Y
Elgiloy®	15% Ni, 20% Cr, 40% Co, 7% Mo, 16% Fe, 2% Mn, 0.1% C, 0.04% Be
Haynes® 188	22% Ni, 22% Cr, 39.2% Co, 14% W, 3% Fe, 0.1% C
L-605	10% Ni, 20% Cr, 52.9% Co, 15% W, 0.05% C
MAR-M 918	20% Ni, 20% Cr, 52.5% Co, 7.5% Ta, 0.05% C, 0.1% Zr
MP35N®	35% Ni, 20% Cr, 35% Co, 10% Mo
MP159	25.5% Ni, 19% Cr, 35.7% Co, 7% Mo, 0.6% Nb, 0.2% Al, 9% Fe, 3% Ti
Stellite® 6B	3% Ni, 30% Cr, 52% Co, 1.5% Mo, 4.5% W, 3% Fe, 2% Mn, 2% Si, 1.1% C
Haynes® 150	28% Cr, 50.5% Co, 20% Fe, 0.75% Si, 0.02% P, 0.002% S
S-816	20% Ni, 20% Cr, Co (bal), 4% Mo, 4% W, 4% Nb, 3% Fe, 1.2% Mn, 0.4% C
V-36	20% Ni, 25% Cr, Co (bal), 4% Mo, 2.3% Nb, 2.4% Fe, 1% Mn, 0.32% C

4.5 Characteristics of Superalloys

4.5.1 Tensile and Yield Properties

In superalloys, the tensile properties are always temperature dependent. Related properties include yield and tensile strength, ductility, and elasticity. One of the special forms of tensile property important for superalloys is the stress rupture strength or the failure of component under static load at elevated temperatures and special environment.

4.5.2 Creep Resistance

Creep is a permanent deformation when the material is exposed to high stresses, which are still lower than its yield strength, for a long period of time. This phenomenon is more pronounced in the presence of high temperature. Under stress and high temperature, gas turbine parts such as disks, bolts, and blades are vulnerable to creep damage; therefore, creep resistance is a critical parameter for turbine parts. Three conditions are needed for creep damage to occur: time, temperature, and stress.

4.5.3 Fatigue Resistance

Fatigue is a type of failure that causes crack or complete breakdown of components under cyclic stress below ultimate strength. The cyclic stresses may not only be mechanical but also thermal or a combined mechanical and thermal. Fatigue resistance is an important design consideration for aerospace component to prevent thermomechanical fatigue due to exposure to extremely fluctuating temperatures.

4.5.4 Corrosion Resistance

Corrosion is a complex phenomenon in superalloys. Corrosion resistance is the material's ability to resist chemical degradation in a specific corrosive environment. Many corrosion-related failure mechanisms exist in superalloys including stress corrosion cracking, intergranular attack, intergranular oxidation, high-temperature oxidation, stress-assisted grain boundary oxidation, hot corrosion [5].

4.6 Industrial Applications of Superalloys

As previously mentioned, development of gas turbines and jet engines was the main driving force in the evolution of superalloys. Operating at high temperatures makes new gas turbines and jet engines more efficient and less pollutant since the combustion cycle is more complete. However, operating at elevated temperatures has its own complications. The main goal is to prudently select materials that can resist very high temperatures, close to their melting point, as well as extreme corrosion and material degradation during their service life. Although superalloys are used in civil applications such as surgical implants and even cutleries, their main field of implementation is high-temperature applications including aerospace components, chemical and petrochemical equipment, nuclear reactors, and heat exchangers [29, 33]. This section is mainly focused on the main application of superalloys in gas turbines and jet engines. Figure 4.4 shows the major consumers of superalloys in industry. It is obvious that more than two-thirds (70%) of superalloys consumption belongs to aerospace industry [34, 35].

4.6.1 Application of Superalloys in Gas Turbines and Jet Engines

The ability of superalloys to withstand intense operating environments, where tensile strength and high-temperature creep, corrosion and oxidation resistance are critical, has made them materials of choice for gas turbine and jet engines that are known to operate at very high temperatures [29, 36]. A variety of other factors such as economic feasibility, service efficiency, and overall performance also influence the popularity of superalloys in gas turbine engines. Similar to other mechanisms and machines, gas turbines and jet engines consist of several components each of which must possess certain characteristics for safe and reliable performance. Superalloys are mainly utilized to build stationary parts such as combustor cans, nozzles, guide vanes, seals, and casings as well as rotating parts including disks,

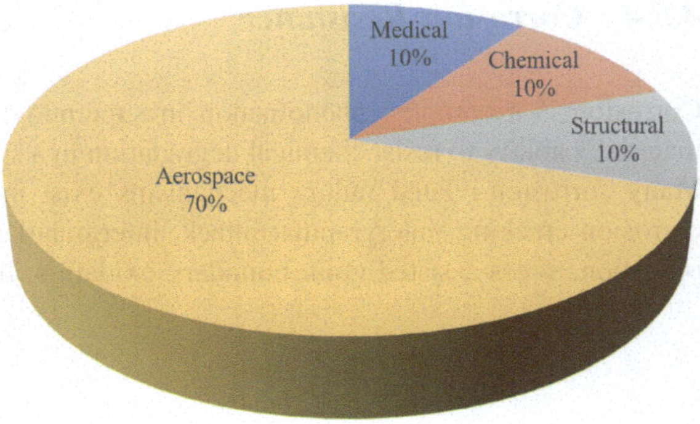

Fig. 4.4 Main consumers of superalloys in industry (with permission to reuse) [34, 35]

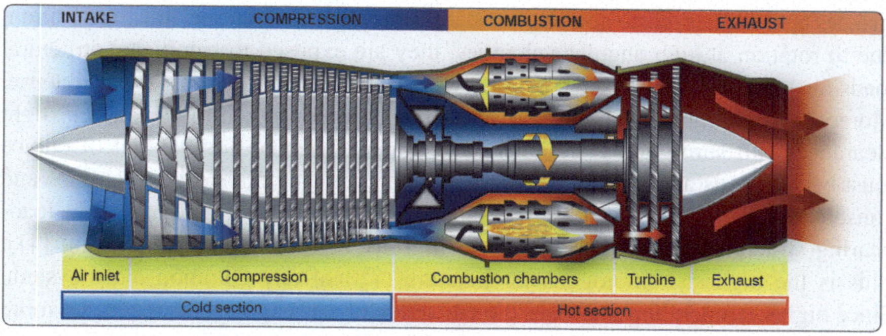

Fig. 4.5 Schematic representation of a gas turbine engine [37]

shafts, blades, and spacers [8]. Figure 4.5 illustrates a typical gas turbine engine, its main components, and their level of exposure to heat.

Among the illustrated components in Fig. 4.5, combustion chambers, where the fuel is ignited, do not carry significant structural load but experience the highest temperatures. Therefore, properties such as creep and oxidation resistance are of primary concerns in the process of material selection for these components [8, 29]. Moreover, since these components are geometrically complex, the materials of choice must have excellent formability and weldability for ease of manufacture. Although some nickel-based superalloys like Nimonic75, Hastelloy X and Inconel 600 can be utilized for the production of combustion chambers [8], cobalt-based superalloys, e.g., Haynes 188, are recommended materials of choice for this particular application, since they possess properties such as higher temperature fatigue and creep resistance. Cobalt-based superalloys are also utilized in the production of turbine vanes as well as other static non-rotating components [15]. These components experience slightly lower temperatures than combustion chambers, but oxidation resistance is still a major concern. Turbine vanes are also exposed to stresses caused by different pressures on the airfoil surface, which make creep-resistant cobalt-based superalloys a desired material. Furthermore, cobalt-based superalloys exhibit better thermal conductivity and lower coefficient of thermal expansion than those of their nickel counterparts. These features make them the primary options for applications where thermal fatigue is the significant design consideration [38]. Cobalt-based superalloys are also widely used as corrosion-resistant and high-strength material in petrochemical and automotive industries. They are also employed is applications such as machine tools, cutlery, and wear-resistant coatings [39].

Another application of superalloys where they demonstrate their superior characteristics is the moving components of gas turbine engines' disks, seals, and blades among many other structures that may experience high temperatures and require high strength [40]. Turbine blades are examples of gas turbine components that are not stationary. Blades are rotating components that are simultaneously exposed to a hot stream of gas and structural loads. Although the temperature they experience is not as high as combustion chambers, it is still high enough to impose thermal stresses and initiate corrosion and oxidation during their service life. In addition, due to rotation at high angular velocities, they are exposed to significant structural loads in terms of centrifugal stresses, which make the situation further intense. Moreover, to maintain high efficiency, the blades are designed to have a very tight clearance with surrounding stationary components; hence, the material of choice must be very resistant to creep deformation to prevent any physical interference and consequent failure [8]. Nickel-based superalloys are mainly implemented in load-bearing structures up to temperatures as high as 90% of their melting point [41]. This is the highest homologous temperature (T_m) of any common alloy system. Since higher temperature increases the efficiency of aeroengines, designers are more interested in the implementation of highly alloyed superalloys, for instance, Waspaloy, which is a nickel-based superalloy demonstrating exceptional fatigue, and high-temperature creep resistance. However, despite its superior characteristics, implementing Waspaloy presents several complications in production and manufacturing [40]. Another nickel-based superalloy in aeroengine industry is Udimet 720 with high-strength/high-temperature potential, which was initially developed as a blade alloy but can also be used to produce large disks. Other nickel-based superalloys in turbine blade productions are Udimet 500, Inconel 100, Udimet 700, and Inconel 738 [8].

An array of different superalloy admixtures based on nickel such as alloyed precipitation-hardened material and dilute solid solution strengthen alloys is implemented in gas turbine engines because they can adequately perform in high temperatures [36]. Transition ducts and combustor liners are examples of complex pieces produced from diluted nickel alloys including IN625 Nimonic 75 and Haynes 230 [36].

Similar to turbine blades, turbine disks are among the moving components of gas turbine engines. Disks are thick circular components whose outer circumference is called the rim, the inner region is known as bore, and the intermediate area between rim and bore is called the web [24]. Bore is the location through which disk is connected to the engine shaft. Turbine disks experience lower temperature than turbine blades; however, the operating condition through its cross section is changing as such the material requirements. Among the disk cross-sectional areas, bore, located at the center, experiences the highest stresses but lower temperatures, which makes the strength and fatigue resistance important considerations. Conversely, the rim experiences higher temperature but less structural load which makes creep a major concern [24]. This results in difficulties in the selection of material with appropriate characteristics to satisfy the demands of each section. Figure 4.6 depicts a schematic radial cross section of a turbine disk with the variation of temperature as well as design considerations [24]. As can be seen, where temperature is lower than 500 °C, strength and fatigue resistance are dominant design factors while for the sections exposed to temperatures above 650 °C creep resistance must be given priority.

N18 is another nickel-based powder metallurgy superalloy with excellent characteristics that designate it as a high-quality material for turbine disks. Turbine disks usually require material with high yield strength in temperatures up to 750 °C,

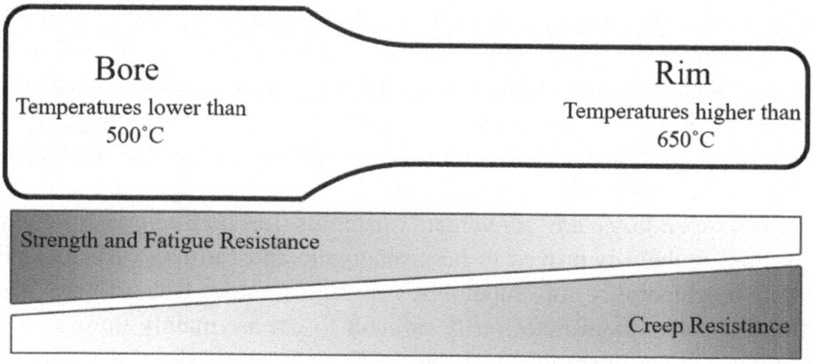

Fig. 4.6 Variation of temperature and key design considerations along a schematic radial cross section of a turbine disk

as well as high creep resistance and damage tolerance up to 650 and 700 °C for long-term and short-term use, respectively [42]. N18 maintains a high ultimate tensile strength (>1500 MPa) up to the temperature of 550 °C and demonstrates a fairly consistent yield strength of 1050 MPa up to the temperature of 700 °C. Additionally, N18 exhibits creep resistance even at 750 °C, which is phenomenal [42].

Other superalloys such as Incoloy 901, Waspaloy, Astroloy, A-286, and Inconel 100 are also used in manufacturing turbine disks. To decrease pressure losses and increase turbine efficiency, clearances are mainly selected very tightly. Components with tight clearances are prone to interference when exposed to high temperatures primarily due to thermal expansion. As a result, application of superalloys with low thermal expansion, for parts with critical clearance like casings, is necessary. The superalloys that possess this characteristic include Incoloy alloys 903, 907, 909, and (more recently) Inconel 783 [8].

4.7 Challenges in the Machining of Superalloys

Machining is widely used in industry to produce parts of different sizes and geometries for divers range of industrial applications. During its manufacturing process, almost every component undergoes some sort of finish or rough machining operations regardless of size, application, and material. Superalloys are of no exception. Among manufacturing operations, machining operations are one of the costliest and most time-consuming ones. In machining, the removed material in the form of chips is discarded and considered as waste; hence, machining must be kept minimized. Based on this fact, industrial trend is more toward near net shaping process such as precision casting, precision forging, and powder metallurgy processing as faster and more cost-effective alternatives [3]. In spite of the recent developments in near net shape processing, machining operations are still very favorable in manufacturing industry due to their flexibility in dealing with different materials and their capabilities in generating complex shapes with high dimensional accuracy and surface quality. Even though it must be kept minimized, machining is still one of the important processes for making parts out of superalloys due to the tight tolerance requirements and geometric complexity in the production of aerospace components such as jet engines parts.

Although superalloys are convenient materials for high-temperature applications, machining them is proved to be problematic and causes substantial difficulties. Poor machinability of superalloys is foreseeable because the material properties that make them notoriously difficult to cut are mainly those that render their unique high-temperature resistance. During machining, these properties impose a high magnitude of heat and stress on the tool and force the operators to keep the cutting speed as low as 5–10% of those typically used for steels [3]. Lowering the cutting speed, imposed by inherent characteristics of superalloys, inevitably reduces the production rate, which is absolutely undesirable in a production setting of large scale. Among the three basic families of superalloys, machining nickel-based and cobalt-based ones is commonly more challenging than iron-based ones. Instead, chip-breaking problems can be observed while machining iron-based base alloys, which often require special tool geometries. Figure 4.7

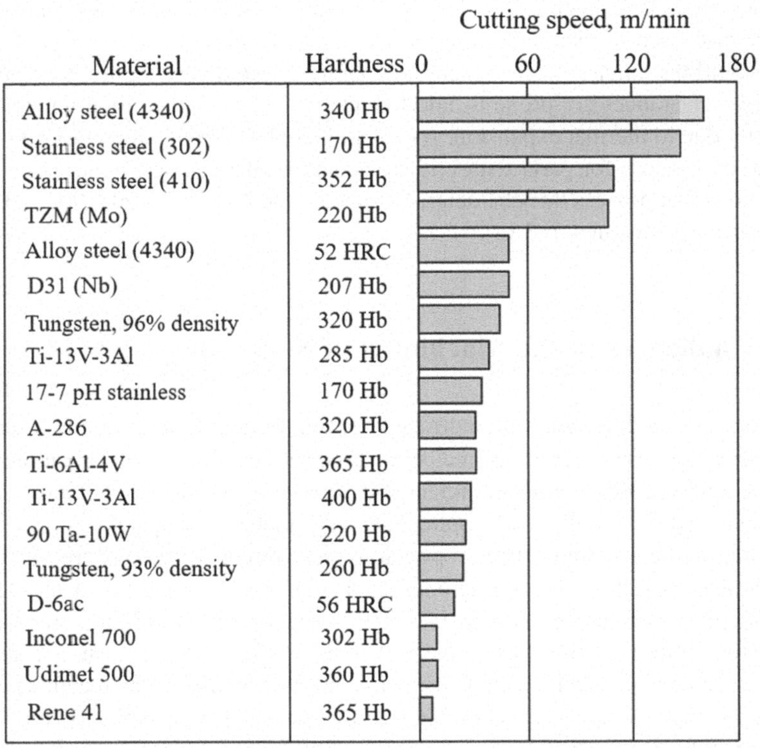

Fig. 4.7 Relative machinability of superalloys versus other materials such as steels, titanium, and refractory metal alloys [3]

shows the typical cutting speed while machining (face milling) of different types of materials including superalloys, stainless steels, titanium, and some refractory metal alloys.

As can be seen from Fig. 4.7, Inconel 700, Udimet 500 and Rene 41, which all belong to the nickel-based family of superalloys, have the least machinability, even worse than that of titanium. Poor machinability of superalloys is mainly due to [3, 43–45]:

- High hot hardness and strength
- High dynamic shear strength
- Low thermal conductivity
- Formation of built-up edge
- Austenitic matrix and subsequent work hardening during machining
- Abrasiveness.

The above-mentioned factors are interrelated, act simultaneously, and contribute to the poor machinability of superalloys. In the following sections, each factor will be discussed.

4.7.1 High Hot Hardness and Strength

Generally, the temperature generated during metal removal process has an obvious effect on the properties of materials. Among the material properties, which are mostly affected by temperature, strength and hardness are inversely proportional to the temperature. On the contrary, ductility is directly proportional to temperature and increases by increasing the temperature. Reduction in strength and hardness is mainly due to increase in the mobility of dislocations. Hot hardness and strength correspond to the ability of material to maintain its hardness and strength at elevated temperatures. Needless to mention here that eventually all of engineering materials lose their hardness and strength at some points when temperature increases; however, superalloys are capable of retaining their hardness and strength up to higher temperatures close to their melting point. As a result of their hot hardness and strength, the elevated temperatures generated during machining do not soften superalloys to any significant degree and therefore do not help improving their machinability.

4.7.2 High Dynamic Shear Strength

As the cutting tool moves toward the workpiece, the stress builds up and material undergoes severe deformation and chip is formed. During the chip formation process, the strain that is built in soft and ductile metals is very large and usually more than 2; hence, the material experiences large plastic deformation [46]. The principal mode of deformation when machining soft and ductile metals is slip. Slip is a deformation process where one part of a crystal past over the other part as a result of the passage of a dislocation through crystal structure. In contrast, when machining difficult-to-cut materials such as titanium alloys or superalloys, specially nickel-based superalloys such as Inconel 718, deformation mode is dominantly twinning [46].

Twinning occurs when shear force deforms the crystal structure such that atoms on one side of a twin boundary are located in mirror position of atoms on the other side [47, 48]. FCC crystal structure and hot hardness and strength, due to the presence of fine precipitates, drastically affect the chip formation during machining superalloys. Examining the deformation area has revealed that material deformation in the chip is completely inhomogeneous. This inhomogeneity, during chip formation, is an indicator of different deformation regions ahead of the tool where each has a different state of stress [46]. When machining superalloys at cutting speeds higher than 61 m/min, the removed layers of material in the form of chips experience low deformation in the bulk of each segments while periodic, localized, intense, thin shear bands are observed between the segments. This shear localization, which is caused by adiabatic shear localization, yields saw-toothed chips with abrasive nature that are very damaging to the cutting tools.

4.7.3 Low Thermal Conductivity

Thermal conductivity is defined as the capability of material to conduct heat. High thermal conductivity facilitates the heat dissipation while low thermal conductivity resists the heat flow. Superalloys have very low thermal conductivity that prevents the effective dissipation of heat from the cutting zone. Superalloys are also very strong even at high temperatures that require higher cutting forces to cut them. Therefore, an extreme amount of heat is produced during machining superalloys that cannot be effectively dissipated due to poor thermal conductivity. Hence, the cutting temperature is localized at the tool tip and reaches more than 1000 °C in some cases [43, 49]. The cutting temperature range of 760–1010 °C has been reported during machining superalloys [3].

Although, in some cases, heat softens the workpiece material to some extent and eases the cutting process, it is normally considered as a detrimental factor in conventional machining processes. Extreme heat, generated during the machining processes, damages the tools and shortens the tool life. At high temperatures, especially those observed when machining superalloys, most of the cutting tool materials experience thermal softening that eventually lead to severe plastic deformation of cutting edge. High temperature is also a contributing factor to the material degradation of cutting tool as it encourages oxidation and diffusion [3].

In addition to plastic deformation and material degradation, excessive tool wear is also a common problem due to the buildup of heat, which is triggered by minimal heat dispersion as well as the high cutting forces caused by excessive friction. One may expect that the high heat experienced by the tool when machining superalloys is a relatively simple problem to fix by lowering the cutting speed and adjusting the feed rate accordingly. This, at the first glance, may reduce the heat and allows for more favorable conditions; however, it significantly increases the production time and consequently decreases the production rate. For a single component, this is of no real concern, whereas for a mass production setting this can have drastic consequences. As a result, heat must be effectively dissipated from the cutting zone to prolong the tool life.

4.7.4 Formation of Built-up Edge

Another common difficulty caused by low thermal conductivity and high friction at tool–chip interface is the formation of built-up edge. At elevated temperatures generated during machining, superalloys tend to adhere to the tool tip and make local welds that are referred to as built-up edges. The size of built-up edge gradually grows by the accumulation of more material until it reaches a certain state at which it becomes unstable and can no longer stay at the tool tip. It is then removed by the flow of chip and another one starts to build up. Each time a built-up edge is removed from the tool tip, a small portion of the tool material or tool coating is also removed. Cyclic formation and removal of built-up edge gradually deteriorates the tool surface and accelerates the wear rate by exposing the substrate unprotected. Excessive tool wear not only makes the process economically infeasible, but also accelerates the frequency of built-up edge formation, which in turn further worsens the situation.

4.7.5 Austenitic Matrix and Work Hardening During Machining

Another feature that negatively affects the machinability of superalloys is their work hardening during machining. Work hardening properties of superalloys mainly originate from their austenitic structure. It occurs when the workpiece material experience severe plastic deformation ahead of the cutting tool. The plastically deformed layer is work hardened and resists further penetration in the subsequent passes [50]. In addition, during the machining of superalloys, the thermal softening induced by high temperature is countered by the strain hardening mechanism, which makes formation of chip even harder. Work hardening and high hot hardness/strength increase the cutting force required to shear the material. Most of this extra force is then converted to heat energy, which due to the poor thermal conductivity of superalloys stays in the cutting zone instead of dissipation to the surrounding area. The complex interaction of these factors increases the tool wear rate and makes machining superalloys a challenging task. The austenitic matrix, especially in nickel-based superalloys, is one of the major causes of severe tool wear and poor machinability of superalloys [43].

4.7.6 Abrasiveness

In addition to all the above-mentioned factors, superalloys also contain abrasive particles [34, 43, 51] due to the presence of alloying elements in their microstructure. These alloying elements are mainly metals that form hard abrasive metal carbide particles (i.e., MC and $M_{23}C_6$). For instance, the age hardening nickel-based superalloys contain abrasive titanium and aluminum particles [50]. The presence of hard abrasive particles in superalloys induces abrasion-related modes of tool wear during machining and worsens their machinability. Moreover, these hard particles are unable to plastically deform during machining and are removed entirely from workpiece surface leaving behind some cavities, which are detrimental for surface integrity [52]. The removed hard carbide particle is then either smeared to the workpiece again or trapped into the chip and slides over the rake face of the tool. The former causes surface integrity problems while the latter severely damages the cutting tool and accelerates tool wear. Rapid tool wear not only makes the machining process uneconomical, but also changes the geometry of cutting tool and increases the cutting forces and creates more heat. The higher cutting forces then increase the friction at the tool–chip interface, which in turn magnifies the abrasive nature of the work material.

4.8 Mechanics of Chip Formation in Machining of Superalloys

The mechanism of chip formation and the nature of chips in machining operations govern different aspects of the process and provide invaluable information regarding the magnitude of cutting forces and their variation, integrity of the machined surface, and even possible modes of tool wear. The mechanism of chip formation becomes more important when the material to be cut is hard, expensive or is utilized in critical applications during its service life such as aerospace, nuclear, and petrochemical industries.

As previously mentioned in this chapter (see Sect. 4.7.2), deformation by twinning is a quite significant factor in the machining of superalloys, especially nickel-based superalloys. Twinning mainly takes place when the slip systems are limited or when twinning stress is smaller than that of slip due to increase in critical resolved shear stress [46]. Resolved shear stress is the component of an applied stress, either tensile or compressive, which is projected along a slip plane. The component along the slip plane acts as a shear component. The magnitude of resolved shear stress at which material starts to yield is called critical resolved shear stress, and it is a material property. However, it is not yet clearly known whether a critical resolved shear stress exists for twinning [46, 53].

The inherent properties of superalloys such as their high strength and FCC crystal structure encourage deformation when the material is locally subjected to a stress level lower than shear stress for slip. In such condition, twinning occurs along (111) planes in [112] directions. However, in case stress reaches the shear stress for slip, slip occurs along (111) planes (similar to that of twinning) but in [110] direction. When twinning occurs, grains are reoriented as a result of the rearrangement of atoms, which may facilitate further slips under favorable conditions [46]. This mechanism is the key factor in the localization of twinning, instability in the primary shear zone and consequent shear localization when machining superalloys especially the nickel-based ones. Consequently, when machining superalloys, the mechanism of chip formation generally involves shear instability and intense localized deformation in the primary shear zone that leads to the formation of serrated or shear localized chips [54, 55]. The same mechanism can be observed when machining titanium and its alloys with HCP structure and limited slip systems.

Shear instability can be considered as a thermomechanical response of the material to the cutting conditions under which it is being machined. It results in shear localization, cyclic chip formation, and cyclic variation of cutting and trust forces. The oscillation of forces is high enough to trigger chatter vibrations even in the most stiff cutting systems [54, 56]. Although shear instability during metal removal processes is a typical characteristic of machining difficult-to-machine materials, it is highly dependent on cutting speed variation [54, 56–58]. It has been shown that the cutting speed at which the transition from continuous chips to shear localized serrated ones occurs is varying from one material to the other once [54]. For instance, in the case of nickel-based superalloys, the transition speed is found to be about 61 m/min [46, 54]. Figure 4.8 shows the transformation of continuous chip to intense shear localized one when machining Inconel 718, a nickel-based superalloy.

As can be seen in Fig. 4.8a, at cutting speed 15.25 m/min that is well below the transition speed of 61 m/min, the chip has a continuous ribbon type with inhomogeneous deformation within the grains. The Fig. 4.8a also demonstrates slip and twinning. By increasing the cutting speed to 30.5 m/min, the chip type is still a continuous ribbon type but the inhomogeneity increases as can be observed from the rise in serration along the free surface of the chip; see Fig. 4.8b. When the cutting speed approaches 61 m/min, the chip shows transition from continuous type to shear localized serrated type. Extensive twinning and shear localized band are clearly seen in Fig. 4.8c. Shear localization is further intensified beyond the transition speed in the narrow bands between the segments; see Fig. 4.8d.

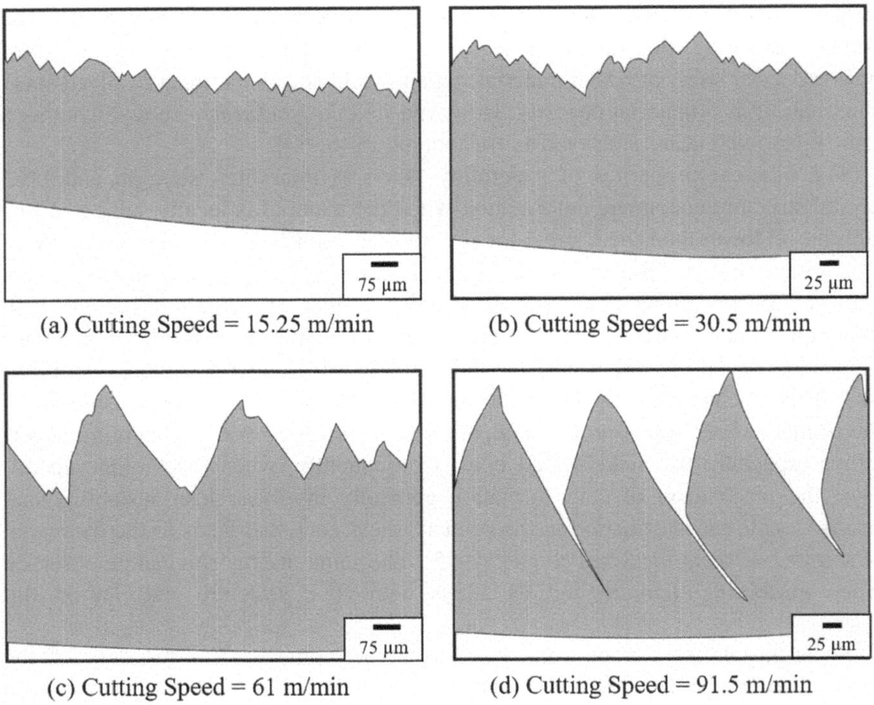

Fig. 4.8 Morphology of chips when machining nickel-based Inconel 718 superalloy at different cutting speed [46]

The mechanism of serrated chip formation in the machining of superalloys, in this case Inconel 718, can be conceptualized based on Fig. 4.9 as follows [55]:

- When the tool proceeds toward the workpiece, the portion of the material located in zone 1 (PA_1A_2), near the tool tip, is exposed to high compressive stresses.
- As cutting tool continues to move forward, this portion is forced upward over the inclined plane A_1A_2. Hence, the material between zone 1 and 2 has a tendency to escape to the free side of the chip and gets pushed toward B_1D_2.
- Concurrently, shearing of workpiece material occurs at an angle ϕ_1 alongside the main shear plane A_1B_1. Then, shear deformation is transformed into heat that softens the chip in the primary shear zone.
- Thermal softening reduces the shear strength of the workpiece material within a narrow band along A_1B_1 and causes a crack initiation and propagation from B_1 to D_2.
- Further movement of the cutting tool produces frictional shear at an angle ϕ_2 along A_2D_2 and the cycle continues.

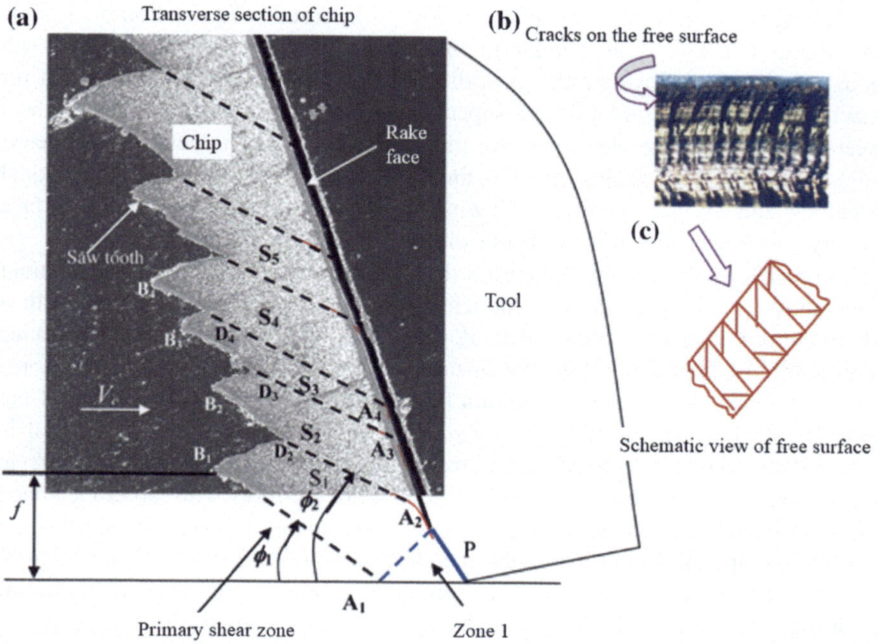

Fig. 4.9 Serrated chip formation in machining of Inconel 718 (with permission to reuse) [55]

4.9 Tool Materials for Conventional Machining of Superalloys

Poor machinability of superalloys imposes several limitations on the effective and economical machining of these materials. Most the promising methods of improving machinability have proved unsuccessful when it comes to machining superalloys. Methods like heat treatment or alloy modification may improve machinability but they adversely affect the unique mechanical properties of these materials; thus, they are considered undesirable. Hot machining is another option of improving the machinability of superalloys; however, it is costly and presents other difficulties particularly when it affects the subsurface deformation and residual stresses [3].

Use of non-traditional machining is considered as an alternative method to overcome the difficulties encountered when cutting superalloys. Several trials were made to combine the traditional and non-traditional machining processes to improve the machining of superalloys; however, non-traditional machining processes have their own drawbacks mainly due to low material removal rate and adverse effects on the surface integrity. For example, implementation of electrochemical machining (ECM) after turning of nickel-based superalloys removes the favorable compressive stresses induced by turning and reduces the fatigue endurance by almost 50% [3].

Among the three categories of superalloys, iron-based solid solution ones show machining behavior quite similar to that of stainless steels, while nickel- and cobalt-based superalloys are the most difficult-to-cut ones. The main approach for machining precipitation-hardened superalloys is to use the solid solution heat treatment to soften the alloy, perform machining to obtain the near net geometry, aging the alloy, and finally run the finishing pass to achieve final dimensional accuracy and surface quality. Following this approach results in good surface quality and longer tool life and better dimensional accuracy [12].

In order to overcome the difficulties that may be encountered during machining superalloys, more specifically minimizing work hardening, a cutting tool with a sharp cutting edge is recommended. Positive rake angles and adequate clearance angles are also preferable [50]. For instance, when machining nickel-based superalloys, the rake angle is selected within the range of 0°–10°. The lower side of the range (0°–5°) is usually preferred for rough machining, where a stronger tool is required due to larger depth of cut and feed. The upper side of the range (5°–10°) is recommended for finishing touches [59]. Using cutting tools with larger clearance angle (also referred to as relief angle or flank angle) is also preferred when machining superalloys but increasing the clearance angle adversely affects the tool strength and weakens the tool. Therefore, the recommended range of optimum clearance angle for machining nickel-based superalloys is 6°–8° for roughing and 10°–12° for finishing [59].

Sharp or lightly honed cutting edge eliminates or minimizes the formation of built-up edge and thus improves the surface quality when machining superalloys. Needless to mention that despite its advantageous character, sharp cutting edges are fragile and prone to chipping; hence, they are not recommended for rough machining of superalloys [34]. When surface quality and dimensional accuracy is not a concern, using tools with larger nose radius is recommended. Large nose radius distributes the chip load over a larger portion of the cutting edge and prevents localized damage and chipping.

In addition to the geometric aspects of cutting tools, the key influential factors that significantly affect the performance of cutting tool materials when machining superalloys are almost similar to those of machining titanium alloys such as [49, 60]:

- High hardness
- Wear resistance
- Chemical inertness
- Fracture toughness.

In addition to tool material and geometry, the work hardening feature of superalloys is an important factor to consider when machining superalloys as it generates a hard layer on the top of newly machined surface. Feed rate and depth of the cut must be adjusted so that in each pass, tool removes the material below the previously work-hardened layer [50]. Low cutting speeds and rigid machining setup, in terms of tool, work holding, and machine structure also reduce the vibration and improve machinability [12]. In general, to achieve successful machining of superalloys for a given machining process, machine tool, cutting tool, and cutting parameters must be carefully selected to strike a balance between advantages and disadvantages. Among the traditional machining operations, which are typically implemented for cutting superalloys, turning, milling, drilling, and grinding are the most common ones [61]. Disks for gas turbines are dominantly manufactured utilizing turning operation while milling is mainly used in manufacturing jet engine mounts and blades for the compressor of jet engines [50, 61].

4.9.1 Appropriate Cutting Tools for Turning of Superalloys

Cutting tools serve at the frontlines of machining industry and directly affect the productivity of the operation. Without a proper choice of cutting tools, tool life will be short and metal cutting operation will be classified as uneconomical. One of the main important parameters in increasing the productivity of machining operations is cutting speed, but it also has the most detrimental effect on the tool life among all machining parameters. In order to achieve higher cutting speeds without compromising the tool life, harder tool materials with higher hot hardness such as coated carbides, ceramics, and cubic boron nitride must be used [62]. Ceramic tools are much harder than carbides, and they can maintain their hardness at higher temperatures. As a result, they can be used at higher cutting speeds of up to 150–200 m/min even higher; however, the feed rate must be selected lower than that of carbide tools (about 80% of appropriate feed for carbide tools) due to lower toughness of ceramics. When cutting hard superalloys such as nickel-based (wrought and cast) and cobalt-based (cast), cubic boron nitride tools are the cutting tool material of choice [3].

High-speed steel (HSS) tools are one of the toughest tool materials in machining industry, and they are being widely used in different applications. However, HSS tools are not an appropriate option for machining superalloys due to the lack of hot hardness, which causes them to drastically lose their hardness at high temperatures. However, in few applications such as turning with interrupted cuts, general-purpose HSS tools of highly alloyed grades are used instead of carbide tools because of their superior toughness and shock resistance. HSS tools for such applications are mainly T15, M36, or M44 grades that show longer tool life than those of general-purpose grades such as M2 or T1; however, they are more expensive [3].

Carbide tools are generally used for machining heat-resistant alloys. Carbide tool can be used in machining nickel-based superalloys at the speed range of 10–30 m/min [49, 63]. Among the carbide tools, tungsten-based carbides have relatively higher toughness than the other grades and are capable of reaching higher feed rates. They can also be used in interrupted cutting; however, the cutting speed at which they can operate is limited due to poor thermochemical instability [49]. Coating carbide tool increases their strength and wear resistance and makes achieving higher speeds possible. Tungsten carbide–cobalt (WC-Co) grades are normally used for machining nickel-based superalloys at the cutting speed of around 50 m/min. These grades are a cemented mixture of hard tungsten carbides and ductile cobalt in the form of a metal matrix composite. Cobalt particles are embedded in the tungsten matrix to increase toughness and shock resistance. Application of carbide tools in high-speed machining of superalloys is not recommended because they cannot sustain high temperatures and stresses that may be encountered.

Ceramic tools are other candidates for machining superalloys. This family of cutting tools is based on two basic types of ceramics, namely aluminum oxide (Al_2O_3) and silicon nitride (Si_3N_4). Ceramic tools based on aluminum oxide, also known as alumina, exhibit high hardness and high compressive strength. Most importantly, they are chemically stable and inert with respect to iron and nickel at elevated temperatures [64]. However, alumina ceramic tools suffer from lack of fracture toughness and thermal shock resistance. The former can be improved by adding zirconium oxide (ZrO_2) while the latter can be addressed by adding titanium carbide (TiC) or titanium nitride (TiN) [49]. These mixed alumina-based cutting tools can be successfully implemented in machining superalloys, especially nickel-based superalloys at cutting speeds in the range of 120–240 m/min, which is much higher than those that are viable with carbide tools. The characteristics of alumina-based cutting tools can be further improved by adding silicon carbide (SiC) in the form of whiskers or fibers. The silicon carbide particles act as reinforcements and improve the matrix (aluminum oxide) properties by lowering the coefficient of thermal expansion and increasing the high-temperature resistance. Whisker-reinforced alumina ceramics can be used in machining superalloys at the cutting speeds ranging from 200 to 750 m/min [49]. The silicon nitride ceramic tools, also known as sialon, demonstrate higher toughness and better thermal properties than those of aluminum oxide ceramic tool, which make them a great candidate in machining superalloys. Compared to alumina and mixed alumina ceramic tools, silicon nitride tools can achieve higher cutting speeds and feed rates when machining nickel-based superalloys. They are also used in the rough machining of nickel–iron-based superalloys [49].

Cubic boron nitride tools, the second hardest cutting tool next to diamond tool, have also been used in machining superalloys, especially in turning Inconel 718 at cutting speeds varying between 120 and 240 m/min. These tools are typically employed in machining nickel-based and cobalt-based superalloys when the workpiece hardness exceeds 340 HV [49].

As superalloys are severely work hardened during machining, which increases the cutting forces, the temperature and rate of tool wear, the single-point cutting tools used in the turning of superalloys must have a positive rake angle to promote cutting rather than pushing the material. Positive rake angle not only reduces the work hardening but also facilitates effective chip evacuation from the finished surface and results in better surface finish [50]. Side cutting edge angle is also an important parameter to be optimized. The objective of such an optimization must be achieving a balance between large side cutting edge angle to provide enough clearance and small angles to maintain strength [50].

4.9.2 Appropriate Cutting Tools for Milling of Superalloys

Due to the rotary nature of milling operation as well as the mechanism of chip formation in milling, tools, fixtures, and machines must be rigid enough to achieve a successful machining of superalloys. Due to the interrupted nature of milling operation and complicated shape and geometry of milling tools, the cutting tool material of choice is normally HSS. HSS possesses good formability to produce complex shapes as well as high toughness and shock resistance to sustain the interruptions during milling operation. However, again, HSS tools suffer from lack of hot hardness that make them susceptible to wear and failure when machining superalloys. Hence, for machining highly alloyed precipitation-hardening grades of superalloys, carbide tools are feasible and recommended options [50, 65].

4.9.3 Modes of Tool Wear When Machining Superalloys

Mechanisms and modes of tool wear when machining superalloys are governed by the characteristics of superalloys as well as the properties of the cutting tools. From the workpiece perspective, superalloys are poor thermal conductors and cannot effectively dissipate the heat through the bulk of material. In addition, they are being work hardened during machining, which in turn increases the magnitude of cutting forces and generates more heat. The generated heat can lead to increase temperature up to 1000 °C when machining Inconel 718 [59]. Both poor thermal conductivity and work hardening negatively affect the tool life. It has been shown that when machining superalloys, the tool experiences a much higher temperature than temperature may be experienced when machining steels. In the case of steels, the heat is dissipated through the bulk of the tool; hence, the maximum temperature is observed somewhere over the rake face back from the cutting edge; while when machining superalloys, the heat is intensified near the cutting edge. As a result, cutting edge and its very close vicinity are the areas of maximum temperature [66]. Consequently, they are also more prone to diffusion wear.

High temperature is proven to be the prime factor for the failure of carbide tools when cutting speed exceeds 30 m/min. In such situation, the cobalt binder phase of carbide tools undergoes thermal softening and the cutting edge fails because of severe plastic deformation [49]. In addition to thermal softening and plastic deformation, localized heat very close to the cutting edge originates crater wear in the very close proximity of the cutting edge, which further weakens the tool and accelerates failure. High compressive stresses along with elevated temperatures also generate bonds between chip and cutting tool and form built-up edge. The separation of built-up edge peels off the carbide tool's coating, deteriorates the substrate, and consequently shortens the tool life. Carbide tools can also be affected by diffusion or solution wear when machining superalloys due to intense temperatures. In this case, the cobalt content of the carbide tool migrates from the tool material

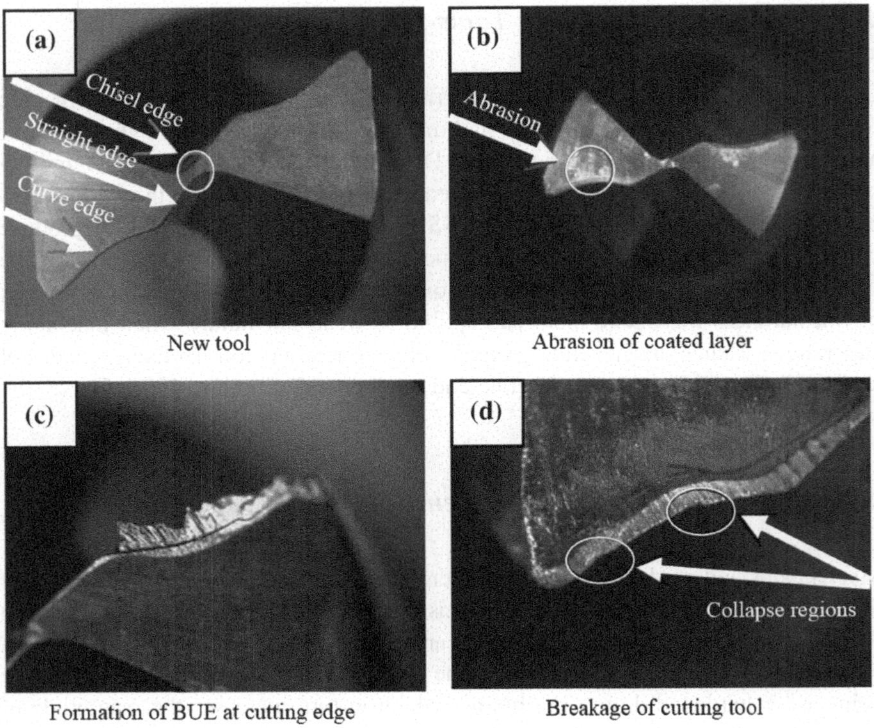

Fig. 4.10 Wear of a TiAlN PVD-coated tungsten carbide twist drill when drilling Inconel 718 at the cutting speed of 400 rpm (13.2 m/min) and feed of 40 mm/min (0.1 mm/rev) (with permission to reuse) [67]

and diffuses into the moving chip; thus, it gradually changes the chemical composition of the cutting tool and makes it susceptible to failure. Figure 4.10 shows the wear progression and failure of a TiAlN PVD-coated tungsten carbide twist drill when drilling a hole on a solid solution Inconel 718. As can be seen, the coating is initially abraded-off followed by the formation of built-up edge and progression of flank wear. Once the tool is weakened enough, chipping occurs at the outer cutting edges. Combined together, these factors lead to tool failure.

Ceramic tools are, however, much harder than their carbide counterparts and maintain their hardness at elevated temperatures. Nevertheless, they suffer from a lack of fracture toughness and flexural strength; therefore, their modes of wear are different. These characteristics make ceramic tools prone to wear in the boundaries coupled with the formation of cracks and notching [49, 59]. Wear of ceramic tools is further accelerated when, due to high temperature, a layer of workpiece (either chip over the rake face or newly machined surface over the flank face) is adhered to the tool. In the presence of high cutting pressures, the adhered layer is squeezed into the longitudinal and lateral cracks on the cutting edge and it makes them extended or wider, respectively. Figure 4.11 shows the modes of tool wear for different

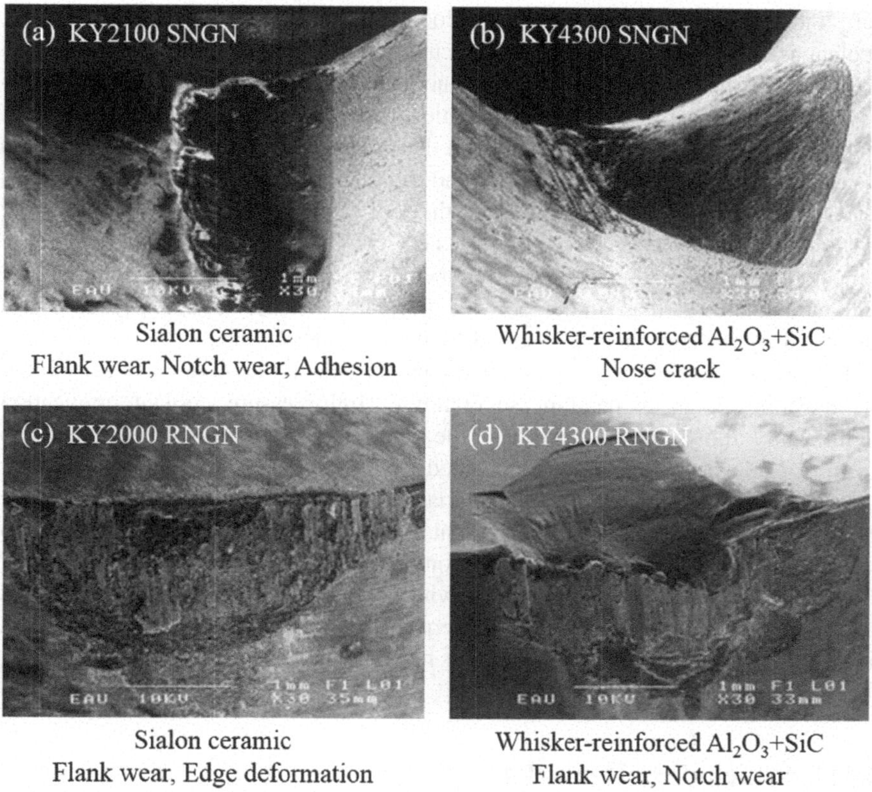

Fig. 4.11 Wear of ceramic tools of different types when turning Inconel 718 at the cutting speed of 300 m/min (with permission to reuse) [60]

ceramic tools when turning Inconel 718 at the cutting speed of 300 m/min. It is clear that flank wear, notching, and nose crack are the dominant mechanism of wear and failure for these tools.

4.10 Application of Coolant in the Machining of Superalloys

Generation of heat is one of the most challenging obstacles when increasing the cutting speed or material removal rate. To increase these two parameters to obtain a satisfactory level for production rate, the excess heat must be removed or eliminated from the system. High-pressure flood cooling is one of the most convenient methods to achieve this objective.

High-pressure cooling condition (HPCC) is a process where a high-pressure jet of coolant is directed at the cutting area. The direct purpose of this procedure is to carry more heat away by simply using a flood coolant. A secondary benefit for using coolant is the lubrication effect of the cutting fluid. In this case, the lubrication properties of the coolant can greatly reduce the amount of heat generated by significantly reducing the coefficient of friction between the tool and the workpiece/chips. Using a high-pressure jet of coolant that is aimed at the cutting face of the tool instead of a simple coolant flood allows more coolant to reach the cutting surface. As well, the coolant jet lifts the chip from the tool face, which leads to reducing the friction and thus heat generation. This effect combined with the reduction in the temperature of the cutting surface allows the cutting speed to be increased up to 50% without a notable increase in the tool wear [68]. The increase in cutting speed is also facilitated by the chip breakage caused by the high-pressure jet.

High-pressure cooling can significantly increase the production rate and viability of machining high-temperature superalloys. High-pressure coolant application significantly allows for an increase in the cutting speed and remarkably increases the tool life when compared to uncoated dry machining. When combined with coated tools, this can lead to significant machinability improvement [69]. However, this process uses large amounts of coolant, which can be expensive and difficult to recycle, and its disposal is also an environmental concern. The implementation of this method requires a complex setup with a large investment and as a result a lengthy return on investment time. Consequently, for low production, this may not be the best option, especially for smaller machine shops.

4.11 Concluding Remarks

Superalloys exhibit superior mechanical, thermal, and chemical properties such as hot strength, hot hardness, high-temperature resistance, high creep resistance, and maintaining chemical and mechanical properties at elevated temperatures. In spite of their excellent properties that make them a primary candidate for high-temperature applications such as components in gas turbine engines or jet engines, machining superalloys always face several difficulties. These difficulties are mainly caused by poor thermal conductivity and high strength and hardness of superalloys at elevated temperatures and strain hardening. Excessive heat interrelated with high cutting forces deteriorate cutting tools rapidly, adversely affect the quality and integrity of workpiece surface, and make the process inefficient. Hence, improving surface integrity, prolonging the tool life, and simultaneously maximizing the production rate are the most important considerations when machining superalloys.

In terms of wear behavior and tool life, mechanical and thermal fatigue, abrasion wear, and diffusion are the main mechanism of tool wear, which result in flank wear and chipping of the cutting tool. When machining superalloys, application of the high-pressure cooling techniques can be very helpful in dissipating the heat, improving the chip segmentation and reducing the thermal fatigue. However, high-pressure cooling increases the cutting edge stresses because it significantly reduces the chip–tool contact length/area. The most efficient cutting tools when machining superalloys, specially the nickel-based ones, are multilayer coated carbide tools, ceramic tools such as sialon and SiC whisker-reinforced alumina ceramics, and CBN tools.

References

1. Reed RC. The superalloys: fundamentals and applications. Cambridge: Cambridge University Press; 2008.
2. Hawk C. Wide gap braze repairs of nickel superalloy gas turbine components. USA: ProQuest Dissertations Publishing; 2008.
3. Donachie MJ, Donachie SJ. Superalloys: A Technical Guide. Ohio: ASM International; 2002.
4. Satyanarayana D, Prasad NE. Nickel-based superalloys. Aerospace materials and material technologies. New York: Springer; 2017. p. 199–228.
5. Hawk C. Wide gap braze repairs of nickel superalloy gas turbine components. Colorado: Colorado School of Mines, Arthur Lakes Library; 2016.
6. Durand-Charre M. The microstructure of superalloys. Boca Raton: CRC Press; 1998.
7. Sims CT. A history of superalloy metallurgy for superalloy metallurgists. Superalloys. 1984;1984:399–419.
8. Geddes B, Leon H, Huang X. Superalloys: alloying and performance. Ohio: Asm International; 2010.
9. Sims CT. Superalloys: genesis and character. Superalloys II–high temperature materials for aerospace and industrial power. New Jersey: Wiley-Interscience, John Wiley and Sons; 1987. p. 3–26.
10. Demmons AC. Superalloy Metallurgy a Gleeble Study of environmental fracture in Inconel 601. San Luis Obispo: California Polytechnic State University; 2016.
11. Pickering F. Metallurgical evolution of stainless steels; 1979.
12. Diltemiz SF, Zhang S. 1 Superalloys for super jobs. Aerospace materials handbook. Boca Raton: CRC Press; 2012. p. 1–76.
13. Holt RT, Wallace W. Impurities and trace elements in nickel-base superalloys. Int Met Rev. 1976;21(1):1–24.
14. Kracke A, Allvac A. Superalloys, the most successful alloy system of modern times-past, present, and future. Superalloy 718 Deriv. 1976;718:13–50.
15. Poole J, Fischer J, Hack G. The development, performance and future of the mechanical alloying process and oxide dispersion strengthened alloys. Advances in high temperature structural materials and protective coatings (A 96–14291 02–23). Ottawa: National Research Council of Canada; 1994. p. 32–53.
16. Coatings N.R.C.C.o., High-temperature oxidation-resistant coatings: coatings for protection from oxidation of superalloys, refractory metals, and graphite. National Academies; 1970.
17. Goward G. Protective coatings for high temperature alloys state of technology. In: Proceedings of the symposium on properties of high temperature alloys with emphasis on environmental effects, vol. 77–1. 1976.
18. Nicholls J. Designing oxidation-resistant coatings. JoM. 2000;52(1):28.
19. Liu R, Yao MX. High-performance wear/corrosion-resistant superalloys. In: Aerospace Materials Handbook; 2012. p. 151.
20. de Lacalle NL, Mentxaka AL. Machine tools for high performance machining. Springer Science and Business Media; 2008.
21. Moniz B. Metallurgy, American Technical Publishers; 1994
22. Committee A. Asm handbook, volume 01-properties and selection: irons, steels, and high-performance alloys. ASM International.

23. Raza SS. Superalloys: an introduction with thermal analysis. J Fundam Appl Sci. 2015;7 (3):364–74.
24. Sharpe HJ. Effect of microstructure on high-temperature mechanical behavior of nickel-base superalloys for turbine disc applications. Georgia Institute of Technology; 2007
25. Sharpe HJ, Saxena A. Effect of microstructure on high-temperature mechanical behavior of nickel-base superalloys for turbine disc applications. In: Advanced Materials Research. Trans Tech Publ; 2011
26. Kushan MC, Uzgur SC, Uzunonat Y, Diltemiz F. Allvac 718 plus™ superalloy for aircraft engine applications. In: Recent Advances in Aircraft Technology. InTech; 2012
27. Davis JR. Nickel, Cobalt, and Their Alloys. ASM international; 2000.
28. Suzuki A, Pollock TM. High-temperature strength and deformation of Γ/Γ′ two-phase Co–Al–W-base alloys. Acta Mater. 2008;56(6):1288–97.
29. Akca E, Gürsel A. A review on superalloys and In718 nickel-based inconel superalloy. Periodicals of Engineering and Natural Sciences (PEN), 2015; 3(1)
30. Hosford WF. Physical metallurgy. Taylor & Francis; 2005
31. Pollock TM, Tin S. Nickel-based superalloys for advanced turbine engines: chemistry, microstructure and properties. J Propul Power. 2006;22(2):361–74.
32. Brandt DA, Warner J. Metallurgy fundamentals. Publisher: The Goodheart-Willcox Company. Inc.; 1985.
33. Sims C, Stoloff N, Hagel WC. Superalloys Ii: high temperature materials for aerospace and industrial power. NY: Wiley; 1987.
34. Ezugwu E. Key improvements in the machining of difficult-to-cut aerospace superalloys. Int J Mach Tools Manuf. 2005;45(12):1353–67.
35. Tools S. Turning difficult-to-machine alloys. Technical Guide; 2002. p. 21.
36. Henderson M, Arrell D, Larsson R, Heobel M, Marchant G. Nickel based superalloy welding practices for industrial gas turbine applications. Sci Technol Weld Joining. 2004;9(1):13–21.
37. Handbook AF. Faa-H-8083-3a. US Department of Transportation Federal Aviation Administration; 2004. p. 1-33.
38. Jovanović MT, Lukić B, Mišković Z, Bobić I, Cvijović I, Dimčić B. Processing and some applications of nickel. Cobalt Titanium-Based Alloys. Metalurgija. 2007;13(2):91–106.
39. Opris C, Liu R, Yao M, Wu X. Development of stellite alloy composites with sintering/hiping technique for wear-resistant applications. Mater Des. 2007;28(2):581–91.
40. Smith R, Lewi G, Yates D. Development and application of nickel alloys in aerospace engineering. Aircr Eng Aerosp Technol. 2001;73(2):138–47.
41. Bowman R. Superalloys: a primer and history. In 9th International Symposium on superalloys; 2000.
42. Guedou J, Lautridou J, Honnorat Y. N18, Powder metallurgy superalloy for disks: development and applications. J Mater Eng Perform. 1993;2(4):551–6.
43. Ezugwu E, Bonney J, Yamane Y. An overview of the machinability of aeroengine alloys. J Mater Process Technol. 2003;134(2):233–53.
44. Sims CT, Hagel W. The Superalloys. Wiley; 1972. **66**: p. 1221–1229.
45. del Prete A, Franchi R, Mariano E. Nickel superalloy components surface integrity control through numerical optimization. In Key Eng Mat. Trans Tech Publ; 2014
46. Komanduri R, Schroeder T. On shear instability in machining a nickel–iron base superalloy. J Eng Indust. 1986;108(2):93–100.
47. Rastani M. Mechanism of slip and twinning; 1992.
48. Callister WD, Rethwisch DG. Materials science and engineering: an introduction, 9th ed. Wiley; 2013
49. Choudhury I, El-Baradie M. Machinability of nickel–base super alloys: a general review. J Mat Process Technol. 1998;77(1):278–84.
50. Li H, Chen X, Zhang S. Zhao D. Tool condition monitoring in machining superalloys. In: Aerospace Materials Handbook. CRC Press; 2012. pp. 77–108.
51. Palakudtewar RK, Gaikwad SV. Dry machining of superalloys: difficulties and remedies

52. Ulutan D, Ozel T. Machining induced surface integrity in titanium and nickel alloys: a review. Int J Mach Tools Manuf. 2011;51(3):250–80.
53. Dieter GE, Bacon DJ. Mechanical metallurgy. vol. 3. New York: McGraw-Hill; 1986
54. Hou ZB, Komanduri R. Modeling of thermomechanical shear instability in machining. Int J Mech Sci. 1997;39(11):1273–314.
55. Pawade R, Joshi SS. Mechanism of chip formation in high-speed turning of Inconel 718. Mach Sci Technol. 2011;15(1):132–52.
56. Komanduri R, Brown R. On the mechanics of chip segmentation in machining. J Eng Indust. 1981;103(1):33–51.
57. Lemaire J, Backofen W. Adiabatic instability in the orthogonal cutting of steel. Metall Trans. 1972;3(2):481–5.
58. Shaw M, Vyas A. Chip formation in the machining of hardened steel. CIRP Ann Manuf Technol. 1993;42(1):29–33.
59. Zhu D, Zhang X, Ding H. Tool wear characteristics in machining of nickel-based superalloys. Int J Mach Tools Manuf. 2013;64:60–77.
60. Altin A, Nalbant M, Taskesen A. The effects of cutting speed on tool wear and tool life when machining Inconel 718 with ceramic tools. Mater Des. 2007;28(9):2518–22.
61. Arunachalam R, Mannan M. Machinability of nickel-based high temperature alloys. Mach Sci Technol. 2000;4(1):127–68.
62. Ezugwu E, Bonney J, Fadare D, Sales W. Machining of nickel-base, Inconel 718, alloy with ceramic tools under finishing conditions with various coolant supply pressures. J Mater Process Technol. 2005;162:609–14.
63. Ezugwu E, Machado A, Pashby I, Wallbank J. The effect of high-pressure coolant supply when machining a heat-resistant nickel-based superalloy. Lubr Eng. 1991;47(9):751–7.
64. Brandt G, Olsson B. Metal powder report conference. Switzerland: Luzern; 1983.
65. Campbell Jr FC. Manufacturing technology for aerospace structural materials. Elsevier; 2011
66. Wright P, Chow J. Deformation characteristics of nickel alloys during machining. J Eng Mater Technol. 1982;104(2):85–93.
67. Chen Y, Liao Y. Study on wear mechanisms in drilling of Inconel 718 superalloy. J Mater Process Technol. 2003;140(1):269–73.
68. Colak O. Investigation on machining performance of Inconel 718 in high pressure cooling conditions. Strojniški vestnik-J Mech Eng. 2012;58(11):683–90.
69. Ezugwu E, Bonney J. Effect of high-pressure coolant supply when machining nickel-base, Inconel 718, alloy with coated carbide tools. J Mater Process Technol. 2004;153:1045–50.

Chapter 5
Metal Matrix Composites

Abstract Composite materials, or in their short-form composites, are a specific category of materials in which two or more materials, as constituents or ingredients, with considerably dissimilar physical or chemical properties are combined together to achieve unique characteristics that might be quite different from those of individual components. In composite materials, the constituent materials remain distinct within the final structure. One of the constituents acts as the main body, which forms the bulk of composite. The main body surrounds and supports the other constituent that usually acts as the strengthening or reinforcing element. Between the two elements, the former is called matrix while the latter is named reinforcement. If the matrix is made of a metallic material, the resultant composite is called metal matrix composite or in short MMC. The reinforcement can be of any metal or other types of materials such as ceramics or organic compounds. Composite materials usually have distinguished physical properties that cannot be found combined in traditional materials. The current chapter explores the composite materials, especially MMCs with the main focus on the challenges that might be encountered during machining of these advanced materials. It presents a brief review of composites' history of evolution, their unique characteristics, and their mechanical properties. Cutting characteristics, appropriate tool materials, modes of tool wear, and other influential factors that must be taken into consideration when machining composite materials will also be presented in this chapter. The chapter ends with an overview of the challenges associated with machining reinforced fiber composites.

5.1 Introduction

The rapid growth of industry continuously challenges the current limitations and boundaries of typical engineering materials and pushes the scientists and engineers to broaden the borders of knowledge to achieve improved mechanical properties. These improved characteristics must satisfy the growing and sometimes conflicting market demands. A clear example of these conflicting demands is in aerospace

H. A. Kishawy and A. Hosseini, *Machining Difficult-to-Cut Materials*, Materials Forming, Machining and Tribology, https://doi.org/10.1007/978-3-319-95966-5_5

industry where high-strength and yet lightweight components are highly needed. It is obvious that to make a part stronger, it must be built thicker, which consequently makes it bulkier and heavier. As a result, building a strong while lightweight part is very arduous, if not impossible, to achieve by conventional methods such as heat treatment or adding alloying elements. Only few materials such as titanium and its alloys are capable of delivering the essential characteristics for aerospace use. However, the use of titanium is hindered by some other factors that are discussed in another chapter of this book. One approach to address the need of industry and build strong and yet lightweight materials is to combine the elements together in a new form. Such a material will demonstrate the desired characteristics and will provide a balance between strength and weight. Composite materials are a new class of advanced materials developed particularly to achieve this goal.

In general, composites are a mixture of several (usually two or more) dissimilar phases that might be chemically and/or physically different [1]. Matrix and fiber are two major building blocks of each composite material. Matrix is the main building block to where the other constituents are added. The reinforcement is the additive part to enhance/reinforce the mechanical properties of the matrix. The composite materials provide far more improved mechanical characteristics than those of its constituents. These significant mechanical properties include but are not limited to high strength-to-density or strength-to-weight ratio (also known as specific strength), high specific stiffness, low coefficient of thermal expansion, and high damping ratio [2]. It must be stated here that careful consideration is due when the material is designed to achieve the above-mentioned desired properties. For instance, types of particles, size and volume fraction are important design parameters in the case of MMCs; while, types of fibers (short or long) and fiber orientations are important design parameters in the case of fiber-reinforced composites.

Nowadays composite materials are widely used in industry. Application of composite materials in different industries varies from simple interior parts of automobiles to very complicated aerospace components [2]. A wide variety of materials, e.g., metals, ceramics, polymers, and glass, can be selected to play the role of matrix or reinforcement. In general, composite materials are categorized based on their matrix materials as follows [3]:

- Polymer matrix composites
- Carbon/carbon, cement, epoxy, and ceramic matrix composites
- Metal matrix composites.

With regard to the above-mentioned composites categorized based on their matrix materials, this chapter is mainly focused on metal matrix composites and their machining characteristics. Metal matrix composites (MMCs) are composed of at least two physically and chemically distinct phases. The two phases are distributed evenly to develop homogeneous properties that are unattainable by any one of the phases individually [4]. This class of composites can be distinguished by their matrix phase (base metal, e.g., aluminum, copper), reinforcement phase (e.g., fibers, particles), or by their manufacturing process (e.g., diffusion bonding, powder metallurgy) [5]. Some examples of MMCs are continuous alumina fibers in aluminum matrix composites, Nb–Ti filaments in a copper matrix, and tungsten carbide (WC)/cobalt (Co) particulate composites [4]. Some advantages of MMCs over the base metals with no reinforcement include higher temperature capabilities and dimensional stability. In comparison to polymer matrix composites, MMCs provide higher strength and stiffness as well as higher electrical and thermal conductivity [4].

MMCs were initially introduced to the aerospace industry in the 1970s, and they found their way to the automotive industry in the late 1980s [6]. MMCs are the primary candidate when mechanical components comprising of high strength-to-weight ratio are highly demanded. This is a common case in automotive and particularly in the aerospace industry where reducing weight while maintaining the strength reduces fuel consumption and consequently saves money [7]. The historical evolution and application of MMCs will be discussed in detail in the following section.

In comparison with other families of composite materials such as ceramic matrix, carbon matrix, or polymer matrix composites, the metal matrix composites offer a number of advantages which include but are not limited to [8]:

- High mechanical strength and enhanced transverse properties
- High stiffness and toughness
- High operational temperature
- Low sensitivity to surface defects
- Good thermal and electrical conductivity.

In addition to the aforementioned advantages, metal matrix composites also provide very good joining characteristics when two components must be connected together. They also have low or no impact on the environment by having no gas emission or moisture absorption issues [8]. Advancement in MMC technology has led to the development of a wide range of microstructures and properties. These properties are, in some cases, far better than those offered by regular metals [5]. Design and engineering of metal matrix composites for particular purposes make it possible to go beyond the boundaries drawn by basic properties/attributes of the main material classes [5]. Significant research on MMCs reached a high level during the late 1980s and early 1990s and continues to date [5].

5.2 Historical Background and Evolution of MMCs

Metal matrix composites have existed for thousands of years, at least as early as 2750 BC, which was evident by an iron laminate that has been found in the Great Pyramid in Giza [9, 10]. However, the evolution history of composite materials, among which the MMCs are one of the most important ones, can be classified into four generations.

5.2.1 First Generation

During an expedition of the Pyramid in 1837, archaeologists discovered an iron plate within one of the unexplored sections of the Pyramid. Upon the analysis of the plate, it was confirmed that the plate's origin was of the same time period of the Pyramid and the plate must have been used in the construction of the Pyramid. The significance of this discovery lays in the fact that this was the oldest plate of iron ever discovered. Secondly, this iron plate was one of the first metal matrix materials that were discovered. Metallographic studies conducted by El-Gayer and Jones revealed that the plate consisted of numerous laminate layers of wrought iron welded and hammered together at temperatures around 800 °C [11]. Further tests showed that these operations introduced inclusions of other non-metallic material into the iron matrix of the plate. Over the course of years, many discoveries were uncovered regarding metal matrix composites from all over the world. Notable examples of these discoveries were Achilles' shield, which was an early form of body armor that consisted of five sheets of bronze/tin/gold/tin/bronze [9, 11]. The complexity and characteristics of the composites continued to improve gradually over the time as new civilizations and technology allowed for upgraded processes and operations.

The first attempt to produce modern metal matrix composites was performed by Eric Schmidt in 1924. His work focused on mixing aluminum and alumina powder together to create a material that was stronger than aluminum. Although his work paved the way for future studies in metal matrix composites, Schmidt was not successful in producing a material that matched the definition of today's metal matrix composite [12]. Throughout the next couple of decades, intensive research was performed and papers were published on metal matrix composites; however, due to the difference between the old and modern definitions, some of the materials that were designated as MMCs are no longer considered as true metal matrix composites [9]. Perhaps, by the 1940s, the first generation in the evolution of composite materials took place but without noticeable developments of metal matrix composites.

5.2.2 Second Generation

The second phase in the development of composite materials, particularly metal matrix composites, took place during the 1960s. It started by the launch of Sputnik in 1957. Sputnik was a Soviet satellite that initiated the "Space Race" between the Soviet Union and the USA. Spacecraft built at that time needed lighter yet stronger materials than the monolithic metals. The desired materials for such applications had to be capable of carrying payload to space while also resisting temperatures up to 1500 °C during reentry stage. A possible reentry nose cone material was developed by the Cincinnati Developmental Laboratories by combining asbestos and phenolic resin. During this period, scientists began to look for a solution through metal matrix composites. Scientists and innovators were inspired by the space race to develop reinforcements such as carbon and boron fibers [9].

Graphite (carbon) fibers were developed and made by Texaco from rayon and boron fibers, which possessed high strength and stiffness properties. Due to its ease of processing, and low cost, carbon fiber was the leading material during that era and it was employed mainly for reinforcing polymer matrix composites not for the metal matrix ones due to the chemical affinity of aluminum and magnesium with carbon. However, the development of air-stable coatings enabled scientists and engineers to prevent the reaction among those metals and carbon. This progress paved the way for the later development of magnesium matrix and aluminum matrix composites reinforced by graphite (carbon) fiber [9]. Boron was mainly used in applications where performance was the primary objective such as military systems. However, implementing boron was challenging due to the following reasons:

- The cost of boron was high.
- The filament could not be deformed or bent to a tight radius.
- Boron had to be deposited over a wire substrate made of tungsten.

In 1966, a contract for development of boron/epoxy was awarded to General Dynamics. The part was aimed to be used as a stabilizer for the F-111 supersonic tactical attack aircraft. Two years later, a contract was awarded to Grumman to develop a composite wing box beam for an advanced fighter [13]. During the second generation in the history of composite materials, the first collected effort to develop MMCs took place. The main objective of that effort was to maintain the high-temperature properties of metallic materials while improving their structural efficiency [9]. Many researchers performed research work on combining metals and reinforcements and explored the utilization of different metals such as iron and nickel as matrices. In addition, several studies investigated the applicability of whiskers and glass fibers as reinforcements. From the perspective of material science, MMCs were not a thoroughly investigated and developed field by that time. Out of all the countries investigating MMCs, the USA was in the forefront where researchers investigated several aspects such as industrial applications and production difficulties in regard to metal matrix composites [9].

5.2.3 *Third Generation*

Around the 1970s and 1980s, material scientists focused on the spacecraft that could be used for multiple space trips. Reusable spacecraft underwent repeated temperature changes when in use, which prompted scientists to look for materials that can handle such conditions. Low coefficient of thermal expansion, resistance to high and cyclic temperatures, and high specific strength altogether were among the most sought-after characteristics for such applications.

During the early 1970s, the knowledge of MMCs was quite undeveloped and immature. Also, due to post-World War II recession, funds for research and development were significantly reduced that ended the early phase of the discovery and development of MMCs. Aggregated by the oil crisis in 1973, the West faced economic concerns that led them to reconsider several energy-related factors like cost and supply. This has led to the realization that energy-efficient materials like composites were a necessity. This realization in turn motivated policy makers to invest more money in composite materials research. The change in the policy was a starting point for resuming research and development of metal matrix composites. The research focus at that time was mainly on boron and carbon fiber-reinforced aluminum matrix composites. Technologies such as scanning electron microscopy (SEM) along with radiography were utilized to study the important characteristics of MMCs. Continuously reinforced metal matrix composites were also investigated with the use of high-strength microfilaments and boron [9]. In the mid-1970s, carbon and boron fibers were used for the first time as the reinforcement in composites with metallic matrix thanks to the development of protective fiber coatings [9].

During the 1980s, research began to focus more on aluminum and titanium matrix composites while also shifting from composites with continuous reinforcement toward those with discontinuous reinforcement. Emphasis on MMCs with discontinuous reinforcement widens the potential range of this type of composite materials. Further research was done during this era on the matrix–reinforcement interface. The main attention was drawn toward interface coating, possible reactions at the interface, and their effect on the properties of MMCs such as fracture toughness and thermal behavior [9]. The processes of powder metallurgy and sintering gained popularity during this period, and plasma technology was used for the first time to produce metal matrix composites. One of the most significant events in the history of MMCs took place in 1986 during US President Ronald Reagan's State of the Union Address. During the address, Reagan acknowledged the presence of the National Aerospace Plane program. The program was about a single-stage aerospace plane, designed to reach space. This plane required the use of materials which must be both very light and very resistant to heat [9]. The only materials at the time thought to possess the attributes to accomplish this task were metal matrix composites since cost was not a design consideration; hence, the influx of research funding was directed to metal matrix composites.

5.2.4 Fourth Generation

During the fourth generation (during the 1990s), the emergence of hybrid materials took place. Hybrid materials were made from the molecular scale combination of organic and inorganic constituents, which turned the scientists' attention toward the possibilities of hybrid structures. During this period, aluminum and titanium metal matrix composites received extra attention from the scientists and engineers. MMCs with continuous and discontinuous reinforcements and their applications at elevated temperatures were the main focus of research. Physical behavior of material such as stress–strain behavior, fracture toughness, and crack growth was also studied. The most important issue with MMCs during this period was residual stresses. The origin of residual stresses, their formation, and the possible remedies were investigated in great detail. Other studies involving metal matrix composites include biaxial mechanics and environmental degradation. Casting, more specifically squeeze casting, was the main manufacturing method for discontinuously reinforced composites with metal matrices.

By the mid-1990s, metal matrix composites were widely used in several aerospace applications including wings and blades, rocket nozzles, and fuselage. Mechanical properties, interfacial properties, stress distribution, FEM modeling, and damage evolution were all examined by researchers during this time. In addition, researchers studied the possibility of heat treatment and the effects of residual stresses on the integrity of composite parts. Development in technology and emergence of the computer allowed for numerical simulation to predict microstructures and stresses. Furthermore, the effects of fiber orientation and particle size as well as aging of metal matrix composites were investigated.

5.3 Characteristics of Metal Matrix Composites

Metal matrix composites have many desired properties that other materials lack thereof. Compared to their metallic counterparts, metal matrix composites are lighter, more dimensionally stable and exhibit higher temperature resistance. They also show better creep resistance and cyclic behavior than metals.

Compared to other composite materials with non-metallic matrix, MMCs exhibit higher thermal conductivity, higher strength and stiffness, and better electrical conductivity. They have better joining characteristics than composites with non-metal matrices. When utilized in extreme service conditions, MMCs receive the least adverse effects from the surrounding environment in terms of contamination or moisture absorption. The superior characteristics of metal matrix composites can be listed as follows:

5.3.1 High-Strength and Improved Transverse Properties

In some applications, high strength is a compulsory property, which may not be satisfied with other types of composite materials such as ceramic matrix or polymer matrix composites. In the majority of composite materials, mechanical strength is a directional characteristic that depends on the alignment of reinforcements. However, in some applications, the mechanical properties must be uniformly distributed through the bulk of material. In these situations, MMCs prove their suitability by providing isotropic properties, in other words by having uniform mechanical strength in every direction.

5.3.2 High Stiffness and Toughness

Generally, reinforcements in composite materials show linear elastic behavior and they do not have an acceptable performance under the impact condition. Mechanical shocks and impact are common cases for the parts exposed to dynamic loading conditions. The desirable impact properties as well as acceptable fracture toughness can be achieved by using composites with ductile metal matrixes. The matrix made of a ductile metal greatly absorbs the dynamic energy during the impact by undergoing plastic deformation.

5.3.3 High Operational Temperature

Despite polymer matrix composites that are very sensitive to high temperatures, metal matrix composites are capable of maintaining their characteristics under much higher operational temperatures due to the nature of their ingredients. MMCs also demonstrate better thermal shock resistance compared to the ceramic matrix composites.

5.3.4 Low Sensitivity to Surface Defects

Metal matrix composites have lower sensitivity to surface defects than other types of composite materials. For instance, polymer matrixes are more sensitive to surface flaws such as small cracks. This is due to several reasons such as low hardness and strength, sensitivity to the surrounding environment moisture, and tendency to react with oxygen (oxidation) in moderate temperatures. Ceramic matrix composites are also very sensitive to surface defects due to the inherent brittle characteristics and high hardness of the matrix.

5.3.5 Good Thermal and Electrical Conductivity

A combination of good thermal and electrical conductivity is a great asset that metal matrix composites offer, which broaden their range of applications. In those applications, where composite parts operate at high temperatures, having good thermal conductivity helps to dissipate the heat and eliminate temperature localization, which may damage the part. Moreover, good electrical conductivity prevents or minimizes the negative effects of lightning strikes by dissipating the electric energy from the impact zone. Incorporating good electrical conductivity prevents or minimizes the negative effects of lightning strikes by dissipating the electric energy from the impact zone. This is a key safety feature in aerospace industry as one of the major destinations for composite materials. The aforementioned characteristics of MMCs accompanied by the availability of an outstanding technical background promote the design, manufacturing, and application of metal matrix composites in various industries. However, some drawbacks challenge the widespread application of MMCs. One of the major drawbacks of MMCs is their tendency to soften and substantial loss of oxidation and corrosion resistance at elevated temperatures. Furthermore, in case of poor chemical compatibility between metal matrix and reinforcement, chemical reactions may occur between the matrix material and the reinforcement, which is undesirable. To prevent the possible chemical reaction between reinforcement and metallic matrix, the reinforcement surface is usually covered by protective coatings.

5.4 Classifications of Metal Matrix Composites

All MMCs contain metals or metallic alloys as their matrix and other metals or ceramic materials as their reinforcement. For this reason, they can be either classified based on the matrix material or the reinforcement. Generally, MMCs are categorized based on the type of their reinforcement [4, 8]. In this section, both classifications will be described.

5.4.1 Classification of MMCs Based on Matrix Materials

A whole array of metals is used to manufacture metal matrix composites. Each metal has its own merits that make it desirable for certain needs and applications.

5.4.1.1 Aluminum Alloys

Aluminum and its alloys are used vastly in automotive and aerospace industries mainly due to their excellent strength, low density, toughness, as well as their ability to resist corrosion. They can be used in the form of cast-, wrought-, or age-hardened alloys [4].

5.4.1.2 Titanium Alloys

Titanium alloys have densities around 4.3–5.1 g/cm^3 and Young's modulus ranging from 80 to 130 GPa that make them important materials of choice for aerospace applications [4]. They have very favorable specific strength (strength-to-density or strength-to-weight ratios) and also possess relatively high melting points. Most of the titanium alloys can retain their strength at high temperatures.

5.4.1.3 Magnesium Alloys

Magnesium has a density of 1.74 g/cm^3, which makes it one of the lightest metals. Magnesium and its alloys, especially the casting ones, can be seen in different industrial sectors. Gearbox housings for automobiles and aircraft, laptop casings, and electronic equipment are few examples, among many, of parts made from this family of alloys. Magnesium alloys show limited plastic formability by slip at room temperature due to their hexagonal close-packed (HCP) crystal structure [4].

5.4.1.4 Cobalt

Cobalt is widely used in tungsten carbide/cobalt (WC/Co) composites as well as cutting tool inserts for machining and oil drills. In this form of metal matrix composites, cobalt in small amounts acts as a binder and keeps the carbide particles in place. Due to its ability to plastically deform, cobalt also provides toughness to the matrix [4]. The composites of this type are mainly formed by sintering.

5.4.1.5 Copper

Copper has a face-centered cubic (FCC) crystal structure. It is ductile and has good thermal conductivity. Moreover, because of its superior electrical conductivity, copper is commonly used as an electrical conductor or as a matrix material in niobium-based superconductors.

5.4.1.6 Silver

Similar to copper, silver has a face-centered cubic (FCC) structure. It is an excellent electrical/thermal conductor with high ductility and corrosion resistance. Lately, it has been used more frequently in high-temperature oxide superconductors to serve as the matrix [4].

5.4.1.7 Nickel

Nickel is another metal with FCC crystal structure that offers good ductility and is used as a matrix in MMCs. Nickel-based superalloys are excellent choices for turbine blades as they have exceptional high-temperature creep resistance.

5.4.1.8 Niobium

Niobium is mainly used to make filaments for composites with superconductivity; however, they are not a very common type of MMCs.

5.4.1.9 Intermetallic Compounds

Intermetallic compounds are formed by chemically combining two dissimilar metals. Ionic and covalent bonds are the main contributing factors for the brittleness and complex crystal structure of intermetallic compounds. When used as a matrix material in metal matrix composites, these compounds tolerate higher operating temperatures over conventional materials.

5.4.2 Classification of MMCs Based on the Type of Reinforcement

MMCs can also be categorized based on the type of their reinforcement as follows [8, 14]:

- Particle-reinforced MMCs
- Discontinues fiber-reinforced MMCs (short fiber or whisker)
- Continuous fiber or sheet-reinforced MMCs.

The last category includes two different types of MMCs that are classified under one title due to their similarities. Table 5.1 illustrates the different types of reinforcements that can be used to reinforce the metallic matrix in MMCs.

Table 5.1 Different types of reinforcements in metal matrix composites (with permission to reuse) [15]

Reinforcement	Aspect ratio	Diameter (μm)	Examples
Particles	1–4	1–25	SiC, Al_2O_3, TiC, B_4C, WC
Discontinues fiber			
(a) Short fibers	10–10,000	1–5	C, SiC, Al_2O_3, (Al_2O_3 + SiO_2)
(b) Whiskers			SiC, TiB_2, Al_2O_3
Continues fiber	>1000	3–150	Al_2O_3, Al_2O_3 + SiO_2, B, C,
			SiC, Si_3N_4, Nb–Ti, Nb_3Sn

5.4.2.1 Particle-Reinforced MMCs

Particle-reinforced composites are the most widely used composites. This is due to the fact that they are the cheapest and easiest to manufacture. Particle-reinforced composites are divided into two subcategories based on the size of particles used as the reinforcement.

In composites reinforced by large particles, the particles act to restrain the movement of the matrix that can only be achieved if the bonding between the matrix and the reinforcement is strong. These composites are used when achieving high levels of wear resistance is required. When the particle reinforcements are used, the resultant MMC shows better stiffness and strength. The properties of particle-reinforced composites are of those of their ingredients. The rule of mixtures is used to create a relationship between the elastic modulus of the matrix and that of the particles.

When very hard and small-sized particles are used to strengthen the metallic matrix, the resultant MMC is called dispersion-strengthened composite. In this type of MMCs, the matrix phase tolerates the bulk of the applied load, and the reinforcement, in the form of very small particles, deals with the dislocation motion. The distribution of the load limits the amount of plastic deformation of the material.

In general, the addition of the reinforcing particles results in isotropic properties across the particle-reinforced metal matrix composites. This isotropic property can be achieved by distributing the reinforcement particles uniformly through the matrix. Mechanical characteristics of particulate-reinforced MMCs mainly depend on the average size of particles, volume fraction, distribution uniformity of the particle in the matrix, particle morphology, and the chemical properties of the particle's surface [2]. In comparison to the fiber-reinforced MMCs, particulate MMCs show inferior anisotropy, greater ductility, and good wear resistance [16, 17].

5.4.2.2 Discontinuous Fiber-Reinforced MMCs

Short fiber-reinforced composites are used when high strength-to-weight ratio is required. A composite material with such a characteristic can be obtained when a soft matrix with low density is reinforced with stiff fibers. The strength of discontinuous fiber-reinforced MMCs is heavily dependent on the length and position of fibers with regard to the direction of the applied stress. The orientation of the fibers is a key factor when creating desired properties. Discontinuous fiber reinforcements are usually in the form of whiskers with an approximate diameter of 0.1–0.5 mm and length-to-diameter ratio of up to 200 [18]. The strength of metal matrix composites reinforced by brittle short fibers is mainly limited by the fiber fracture. In a case where the stress/strain sustained by the composite becomes equal or exceeds the failure limit or allowable strain of the fiber, the composite will fail. MMCs with ductile short fiber do not fracture but undergo yield and plastic deformation in company with the matrix. As well, the strength is still constrained by fiber failure strain [19].

Discontinuous fiber-reinforced composites such as whisker-reinforced ones have uniform properties in all directions and offer a high strength-to-weight ratio. The yield and tensile strength of this type of composites can be improved by increasing the volume fraction of the reinforcement. The tensile modulus of discontinuously reinforced composites is much lower than the modulus for continuous fiber composites. In this case, the type of reinforcement will not affect the modulus. In this case, the strength mainly depends on the type of alloy used in the matrix than the type of reinforcement utilized. In general, the stronger matrix alloy results in a stronger composite.

The whisker size has a large impact on the properties of whisker-reinforced composites such as yield strength, tensile strength, and ductility [19]. The tensile and yield strength have an inverse relationship with whisker size, where they will increase with a decrease in size and vice versa. The ductility of the composite will also increase as the particle size decreases. When using the same material as a reinforcement, a discontinuously reinforced metal matrix composite will be stronger and stiffer than particle-reinforced composite. These metal matrix composites are known for having excellent durability and high corrosion resistance.

5.4.2.3 Continuous Fiber and Sheet-Reinforced MMCs

The fiber-reinforced metal matrix composites are generally made from a ductile matrix with low yield strength reinforced by brittle fibers that have higher mechanical strength. In spite of the particulate-reinforced MMCs, which have isotropic properties, the fiber-reinforced MMCs show directional mechanical properties being maximum along the fibers' direction. The length and type of fibers can vary from short length, which is known as whiskers, to long and continuous ones. The fiber-reinforced MMCs demonstrate comparatively better fracture toughness and temperature resistance than particulate MMCs [20].

Different stress–strain stages can be observed when uniaxial fiber-reinforced MMCs are loaded in the direction of fibers. At the beginning, fibers and matrix deform elastically together until the matrix, which normally has a lower modulus of elasticity, reaches its yield strength. At this stage, the fibers continue to deform elastically while the matrix undergoes plastic deformation. The second stage ends when fibers reach their yield strength after which both matrix and fibers deform plastically. The fourth or last stage takes place when the high-strength part of the MMC, which is fiber, reaches its ultimate tensile strength and the entire composite fails or fractures.

Sheet-reinforced, or in other words laminated, metal matrix composites are generally produced by reinforcing a ductile metallic matrix with repeated lamellar reinforcement of high mechanical strength. The mechanical characteristics of sheet-reinforced MMCs are closely related to those of bulk reinforcement. In comparison to the fiber-reinforced

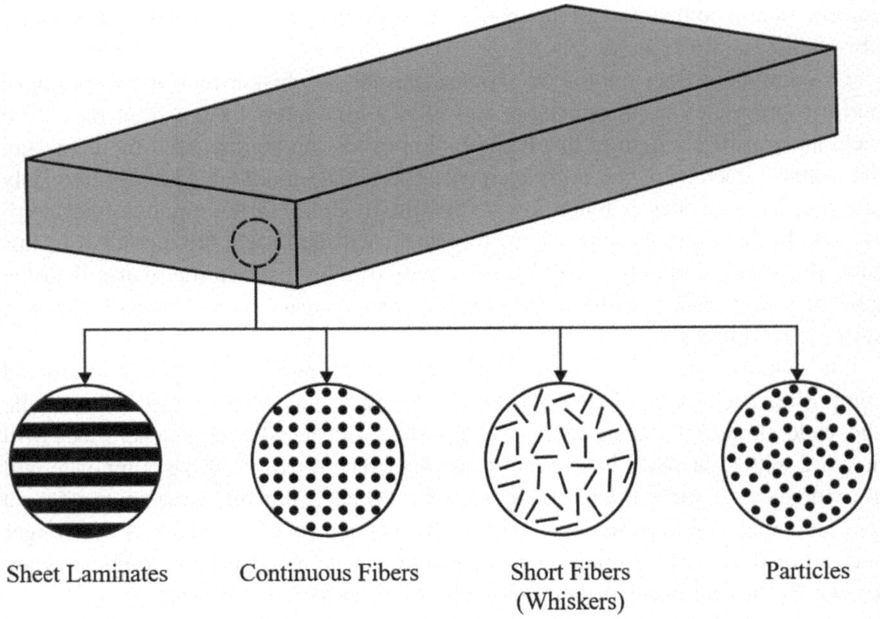

Fig. 5.1 Different types of reinforcements used in metal matrix composites

MMCs, the sheet-reinforced or laminated MMCs show lower ductility, strength, and elongation. Figure 5.1 shows schematically the different types of reinforcements used in typical metal matrix composites.

5.5 Industrial Applications of Metal Matrix Composites

Given the general characteristics of MMCs, it is easy to understand why they are appealing materials for different industrial applications. Generally, metal matrix composites are useful for industrial applications that require materials with high strength-to-weight ratio, good thermal and mechanical properties, and versatility. Industries that utilize the advantageous properties of MMCs include but are not limited to aerospace, automotive electronics and thermal management, filamentary superconducting magnets, and recreational products and sporting goods [4]. In this section, applications of metal matrix composites in aerospace and automotive industry will be discussed.

5.5.1 Aerospace Applications

Metal matrix composites are widely employed in aerospace applications owing to their low density, low coefficient of thermal expansion, high stiffness and strength, and high conductivity. Considering the tight safety requirements in aerospace industry, the service performance of the utilized materials is of much higher importance than the cost considerations.

An example of MMCs' application in aerospace industry is rockets. Space rockets undergo immense quantities of heat, and thus, the materials utilized in such applications must be able to withstand temperatures in excess of 3200 °C [21]. Military aircraft structures are among other examples that require increased strength and specific stiffness to improve the performance of aircraft. For instance, stringers, which are thin strips of material to which the skin of the aircraft is fastened, are aimed to stabilize the skin in the fuselage. Skin buckling is less likely to occur because of the increased skin stiffness [4]. Another example for successful utilization of metal matrix composites in aerospace industry is F-16 Falcon fighter jet in which the aluminum doors used to be vulnerable to fatigue cracking. The issue was addressed by replacing them with reinforced MMC (Al/SiCp) doors without compromising the weight [4]. MMCs were also used to replace the ventral fins because the material showed much lower deflection because of its high stiffness. It helped increase the life cycle of the components, and it ended up saving $26 million in life cycle saving costs [4]. MMCs are utilized in commercial aircraft such as Boeing 777 [4] to replace the carbon/epoxy composite which was originally used to build fan exit guide vane of the engines. The material replacement was triggered by the fact that carbon/epoxy composite was very vulnerable to foreign object damages (FODs).

MMCs are also employed to build blade sleeves in helicopters. Blade sleeve must safely withstand the centrifugal loads from the blades of the rotor [4]. It must also satisfy factors like fatigue life, toughness, and high specific strength. To achieve these properties, silicon carbide (SiC) particles are used in an alloy matrix. Billets can be formed by the use of powder metallurgy and then extruded, cut, and forged. The resultant cost is lower than the original titanium alloy that used to be implemented for that application [4].

Missiles are another important field of application for MMCs in aerospace and defense industry. Conventional aluminum alloys are neither strong enough nor temperature resistant to cope with the demanding service conditions of such applications. Steels or titanium alloys have the needed strength but they cannot be used due to their relatively heavy weight. MMCs are good candidates because they can provide higher strength and stiffness without jeopardizing the weight. They are also exposed to elevated temperatures of the missile for only a short duration that can be well sustained [4].

5.5.2 Automotive and Transportation Applications

In automotive industries, cylinder heads, liners, pistons, brake rotors, and calipers [22–24] are examples where MMCs are used to achieve better wear and thermal resistance. However, the application of MMCs in the automotive industry is not limited to the aforementioned parts.

Automotive and railway industries use MMCs in different varieties of applications. One of the commercial applications of MMCs in automotive industry is the Toyota diesel engine. It had alumina–silica fiber incorporated in the ring groove of the piston during the pressure casting of aluminum [4]. The pistons for diesel engines are usually made of Al–Si casting alloys. The difference between the two is that the MMC exhibited better wear resistance. Compared to aluminum, the MMC also offers a lower coefficient of thermal expansion. This feature reduces the clearance and helps the piston to maintain a better seal with the cylinder and consequently improves the engine performance [4].

The connecting rod, which is made by steel, can be replaced by a lighter aluminum one reinforced by SiC particles. A connecting rod must possess high fatigue resistance at temperatures around 150 °C. Since the Al matrix composites reinforced by SiC particles are lighter than those made of steel, the connecting rod made of MMC would achieve 12–20% reduction in secondary shaking force, 0.5–1% increase in fuel economy, 15–20% increase in maximum RPM, reduced bearing width, and ultimately improved durability of bearing and crankshaft [25].

5.6 Challenges in the Machining of Metal Matrix Composites

As it was previously mentioned, the application of metal matrix composites in numerous industrial areas has rapidly increased. MMCs owe this great business interest to their unique mechanical and physical characteristics. As a result, the materials that are conventionally used in automotive and aerospace industries can be effectively replaced by MMCs to reduce weight while maintaining similar or even improved strength.

Production processes for manufacturing parts out of MMCs are categorized into primary and secondary processes. Primary processes include casting or forming of initial billets, ingots, or blank sheets while secondary processes can be either net shaping like machining or joining operations. Challenges often arise when it comes to machining metal matrix composites. MMCs are categorized as hard-to-cut materials due to the abrasive nature of the reinforcements [20]. Despite the superior characteristics of MMCs, their poor machinability represents challenges, which hinder their utilization in several industries. It has been shown that the cutting tools are rapidly deteriorated while machining MMCs due to the presence of hard reinforcements such as glass, graphite, boron, alumina, and silicon carbide [20, 26].

These reinforcement materials are very abrasive and in some cases harder than the cutting tool materials. The machinability of MMCs in terms of tool life and integrity of machined surface is negatively affected by the presence of these hard particles.

In spite of the recent advancements in manufacturing processes which make the manufacturers capable of producing MMC parts to near net shape, still many conventional machining operations like turning and milling are extensively used to remove the excess materials to produce the final shapes. In addition, to join the composite parts together to form mechanical assemblies, drilling is widely used to generate holes for bolts or other types of fasteners. Considering their potential applications, the associated challenges in the machining of MMCs must be thoroughly investigated to improve the machining efficiency by minimizing the adverse effects of abrasive reinforcements on tool life.

Machining MMCs is greatly influenced by the material characteristics of reinforcement and matrix, type and dimensions of reinforcement (particle or whisker), and volume fraction and distribution of reinforcement through the matrix [27]. Therefore, machining of MMCs can be studied under two main different classifications based on the type and shape of reinforcements as follows:

- Machining of particulate-reinforced metal matrix composites and
- Machining of fiber-reinforced metal matrix composites.

Since fiber-reinforced metal matrix composite parts are normally made by molding processes, they require less machining than the particulate-reinforced MMCs. The most frequently utilized machining operations, in the case of fiber-reinforced MMCs, are drilling and trimming to achieve required dimensional accuracy or making assembly by joining the composite parts together, respectively.

On the other hand, the majority of particulate MMCs are manufactured close to their final shape (near net shape) by forming and casting processes [16]; hence, usually some further traditional machining operations such as turning, milling, and drilling are needed to obtain the final dimensions and the desired surface quality. In contrast, conventional machining operations are rarely used for machining fiber-reinforced MMCs.

5.6.1 Machining of Particulate-Reinforced MMCs

The first and the most important step toward comprehending the mechanics of machining (cutting process) and consequently predicting the cutting forces is studying the mechanics of chip formation. During machining metals or metallic alloys, the cutting tool is continuously in touch with a homogeneous bulk of material; therefore, the tool is exposed to the same material and it experiences homogeneous mechanical properties (see Fig. 5.2).

In machining of homogeneous materials, by assuming a tool with a perfect sharp cutting edge (see dotted line in Fig. 5.2), the chip is formed as a result of shearing

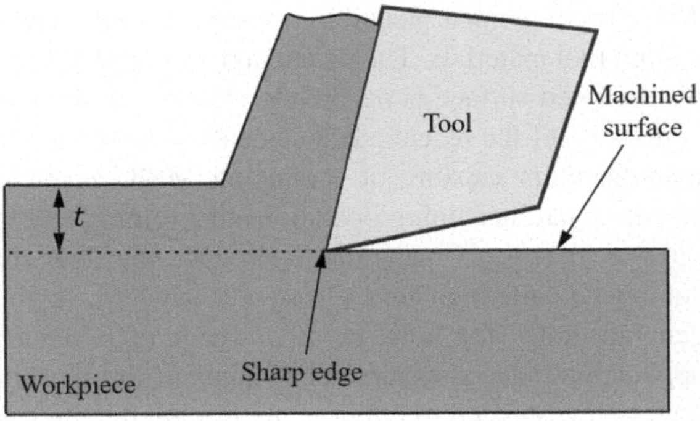

Fig. 5.2 Machining metals or metallic alloys using sharp edge tool

the workpiece material along the shear plane (or thin shear zone) caused by the relative motion of the tool and workpiece. To be more realistic, a perfect sharp cutting edge is almost impossible to achieve due to the limitation of manufacturing processes; hence, cutting tools normally have a round edge with a certain edge radius (see also Fig. 5.2). The actual undeformed chip thickness (t) is changed when the cutting edge is round. In this case, the effective undeformed chip thickness (t) is reduced to ($t - \Delta$) as can be seen in Fig. 5.3.

However, the story is different when machining composite materials, considering their inhomogeneous nature. Machining of composite materials in general and metal matrix composites in particular is closely related to the mechanical and physical characteristics of fiber and matrix. The mechanics of chip formation in machining metal matrix composites slightly differs from that of homogeneous materials. As previously mentioned, the cutting tool is in contact with homogeneous material where the mechanical and physical characteristics are similar in all directions. In the case of machining MMCs, the cutting tool faces randomly distributed reinforcements in the form of fibers or particles during its cutting motion. The presence of these reinforcements, which are much harder than the bulk of material (matrix), changes the plastic deformation behavior of the material. Therefore, the matrix material, which is normally selected from soft metals, exhibits different deformation mechanism than that of the conventional metals or metallic alloys. The chip formation during machining MMCs involves a combination of

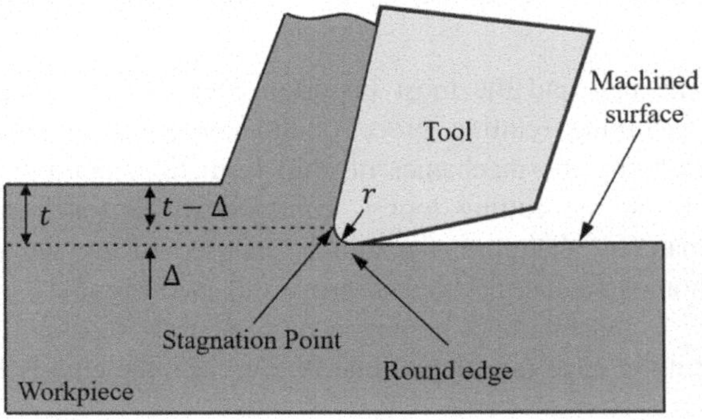

Fig. 5.3 Machining metals or metallic alloys using round edge tool

shearing, ploughing, particle interface debonding as well as particle fracture, fiber pullout and cracking [2, 20]. Figure 5.4 illustrates a schematic diagram of the chip formation during machining MMCs as well as the possible relative positions of the reinforcements and tool tip during cutting.

As can be seen in Fig. 5.4, one of the following three possible scenarios can happen to the reinforcement based on the relative position of the reinforcement and cutting edge [2]. It must be mentioned here that the reinforcement particles are randomly distributed through the body of matrix and the aforementioned scenarios are the only possible situations that can happen to any particle [28].

- The reinforcement particle is located above the cutting line.
- The reinforcement particle is located below the cutting line.
- The reinforcement particle is located along the cutting line.

In the first case, where the particle is located above the cutting line, it will be pushed toward the chip and removed from the workpiece body. In the second case, where the particle is located below the cutting line, it will be pushed further into the body of matrix. The displacement of the particles beneath the machined surface affects the state of stress and alters the generated residual stresses beneath the surface. Due to the difference in material characteristics of matrix and reinforcement, the occurrence of any of these two situations generates dislocations along the interface of the matrix and particle.

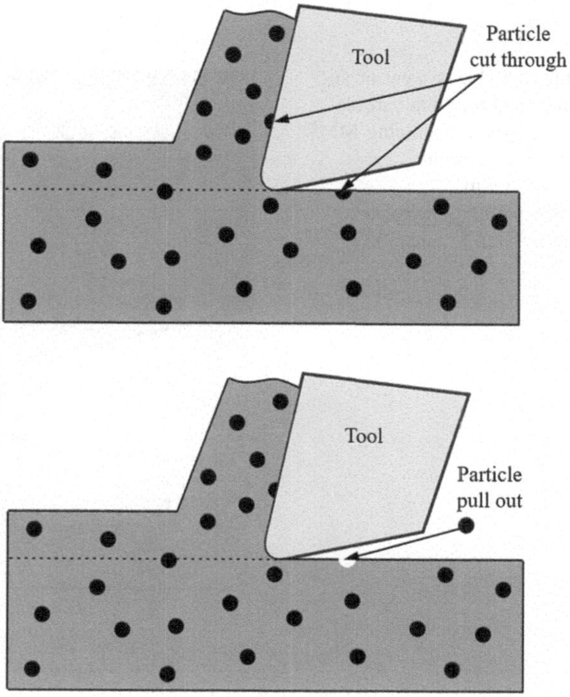

Fig. 5.4 Mechanics of chip formation when machining MMCs and possible options for tool and reinforcement interactions

In the second and third cases, where the reinforcement particle is nearly along the cutting line, it will be either cut through or completely detached from the matrix body. Between these two possible mechanisms, the sheared particle is preferred as it produces better surface finish while a completely pulled up particle leaves a cavity behind and deteriorates the quality of the machined surface. These cavities are advantageous, as they will serve as reservoirs of oil in contact applications. The pulled up particle may also freely roll over either along the tool–chip interface or tool–workpiece interface and would severely damage both tool by wearing and workpiece by damaging surface quality. These different scenarios contribute to the two and three-body abrasive wear observed when examining worn cutting tools.

Figure 5.5 shows a worn-out uncoated tungsten carbide inserts used as a cutting tool for machining 6061 and A356 MMCs reinforced with alumina and silicon carbide (SiC) particles. Similar observations were reported by many researchers [29–31]. Marks of uniform abrasion wear tracks can be seen in Fig. 5.5. Two-body abrasion is indicated by grooves parallel to the cutting direction [32]. These grooves were generated under severe abrasion conditions along the contact between the tool and the reinforced ceramic particles. Three-body abrasion is also observed but its formation is much more complicated than the two-body abrasion. It is formed by the rolling of debonded particles between the cutting tool and the matrix. In general, the sliding of ceramic particles along the tool surface generates two-body abrasion and the free rolling of the abrasive particles leads to three-body abrasion. The indentation marks by the freely rolling particles are shown in the examined tools and indicate the existence of three-body abrasion.

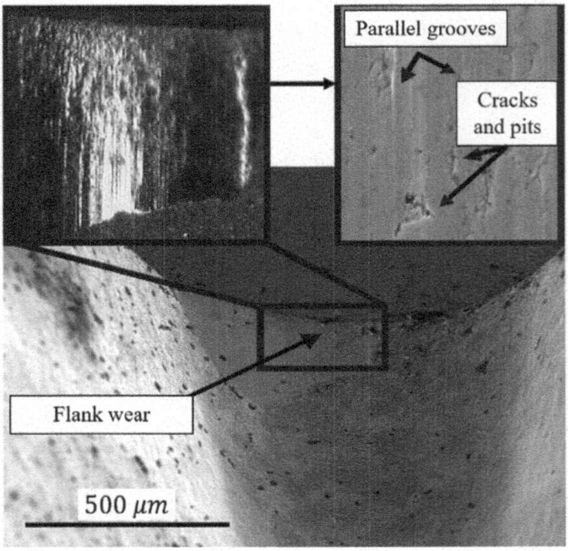

Fig. 5.5 Flank wear of an uncoated tungsten carbide insert when machining MMC (V_c = 185 m/min, f = 0.3 mm, DOC = 0.1 mm) (with permission to reuse) [32]

5.6.1.1 Chip Formation in the Machining of Particulate-Reinforced MMCs

Generally, in metal cutting, the chip is formed when the workpiece material is plastically deformed along the shear zone. The plastic deformation along the shear zone can be defined by two different models [33–35]:

- Thin-plane or thin-zone model and
- Thick-plane or thick-zone model.

The experimental results demonstrate that the material behavior under a very low cutting speed can be described by the thick-zone model while at the higher speeds thin-plane model provides a better approximation. In reality, metal matrix composites are normally machined at moderately higher cutting speeds than other materials. Thus, an acceptable estimation of the cutting process can be achieved by applying the thin-plane model [2, 20]. In addition, the mechanics of chip formation is also affected by some other factors such as mechanical properties of workpiece materials, cutting conditions, and geometry of cutting tool.

The possible types of generated chips during a cutting process can be divided into several categories. Although there is no unique classification for the chip's type, the following classifications are widely accepted [33–35]:

- Continuous chip
- Discontinuous chip
- Built-up edge chip
- Serrated chip.

Among the above-mentioned types of chips, continuous chips are generally observed when the cutting speed is high or the cutting tool has a positive rake angle. Discontinuous chips are usually generated during machining of brittle materials, and built-up edge chips are seen when cutting speed is low. The last category of chips which is the serrated chip is the common type in the machining of materials with low thermal conductivity such as titanium [20]. When it comes to machining metal matrix composites, chips with different shapes and geometries are formed depending on the cutting conditions such as feed and cutting speed as well as material properties [36].

Machining of non-reinforced Al-6061 and MMC made of same alloy reinforced with 20 vol.% SiC particles (size 6–18 μm) revealed that feed is highly influential on the shape of chips when MMC is machined but it does not have a significant effect on the chip shape when non-reinforced alloy is machined [36]. It has been shown that at lower feeds (0.025 mm/rev), chips are mainly short and irregular, while long spiral and straight chips are likely to form at slightly larger feeds (0.05–0.1 mm/rev). Short and C-shaped chips have also been reported by further increasing the feed [36]. Cutting speed has been proved as another influential factor on the chip shape when machining MMCs. Machining above-mentioned MMC at speeds in the range of 100–200 m/min yielded spiral-shaped chips, but at higher cutting speeds in the range of 400–600 m/min, chips of straight shape were formed. Other studies also reported discontinuous chips during machining MMCs at very low or very high cutting speed [20]. Interestingly, long spiral chips are produced when machining non-reinforced alloy regardless of the cutting speed [36]. The variation of chip shape during machining is mainly affected by the shear localization and uniformity of deformation [37]. In the case of MMCs, the presence of reinforced particles causes local deformation and stress concentration. During machining, when the MMC is cut and moves through the primary and secondary shear zone, it experiences severe strain that makes the matrix work-hardened. In addition, some particles are usually debonded during this process that initiate cracks in the matrix. These two factors combined yield brittle and easy-to-fracture chips. This explains the formation of short chips at relatively high feeds. At average values of feeds, chip experiences a more homogeneous deformation across its thickness; thus, longer chips are formed. However, very small feeds result in the formation of very thin chips that are highly vulnerable to breakage and they form short irregular shapes [36]. Same conclusion can be made for the effect of cutting speed on the mechanics of chip formation during machining MMCs at low and high cutting speeds. At lower cutting speeds, inhomogeneous deformation is caused by the dominant effect of strain rate that causes spiral chips to form. On the other hand, at higher cutting speeds, thermal effects on the ductility of matrix are more dominant; as a result, straight chips are likely to form [36]. Figure 5.6 shows the effect of feed on the possible shape of chips when machining MMCs.

Other types of chips such as serrated, discontinuous, and built-up edge chips are also observed during machining MMCs. For instance, built-up edge chips are very

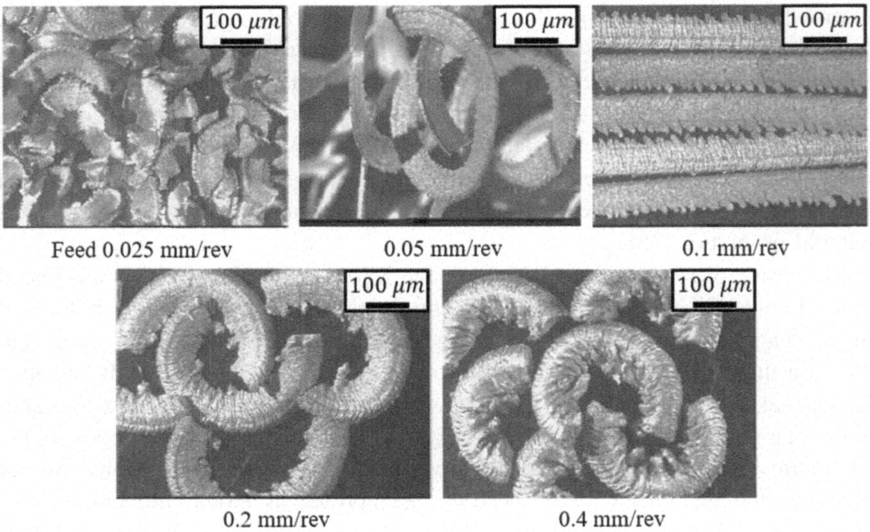

Fig. 5.6 Effects of feed on the chip formation at speed 400 m/min and depth of cut 1 mm (with permission to reuse) [36]

common when aluminum matrix composites are being machined. The geometry of cutting tool is also an important factor in generating chips of certain type in machining MMCs. It has been found that in machining aluminum alloy metal matrix composites, the continuous chips are produced when the cutting tool is new and sharp, while semi-continuous chips are likely to observe when the tool is worn out or the feed and depth of cut are high [16, 38, 39].

5.6.1.2 Cutting Forces in the Machining of Particulate-Reinforced MMCs

When machining particulate-reinforced metal matrix composites, cutting forces are generated by the combination three following mechanisms [40]:

- Shearing of the metallic matrix during chip formation
- Ploughing of the metallic matrix when exposed to rounded cutting edge and
- Fracture or pullout of the reinforcing particles from the matrix.

As a result, the two main components of machining forces, namely cutting force and trust force, can be presented as follows [40]:

$$F_C = F_{CC} + F_{CP} + F_{CD} \tag{5.1}$$

$$F_T = F_{TC} + F_{TP} + F_{TD} \tag{5.2}$$

where F_C is cutting force, F_T is trust force, F_{CC} and F_{TC} are cutting and trust forces produced when cutting the metal matrix, F_{CP} and F_{TP} are the ploughing forces along and normal to the path of cutting tool, and finally F_{CD} and F_{TD} are the forces generated by the debonding and fracture of the reinforcing particles. In order to predict the cutting forces when machining MMCs, each of the above components of force must be accurately calculated. As previously mentioned, due to the relatively high cutting speeds at which MMCs are machined, thin-plane model or shear plane is a good estimation of MMCs behavior during chip formation. According to shear plane model, the cutting and trust forces can be calculated using Eqs. (5.3) and (5.4).

$$F_{CC} = \frac{\left(\frac{1}{\sqrt{3}}\right) kab \cos(\beta - \gamma)}{\sin\phi \cos(\phi + \beta - \gamma)} \tag{5.3}$$

$$F_{TC} = \frac{\left(\frac{1}{\sqrt{3}}\right) kab \sin(\beta - \gamma)}{\sin\phi \cos(\phi + \beta - \gamma)} \tag{5.4}$$

In the above equation, ϕ is shear angle, γ is rake angle, β is friction angle, k is the shear strength of the metal matrix, and finally a and b are the equivalent depth of cut and feed, which can be calculated by implementing the concept of equivalent cutting edge [41]. In order to use this concept, the influence of the cutting tool nose radius must be taken into account by replacing the straight and round portion of the cutting edge with a single straight edge joining the extreme points of engagement [40, 41]. Therefore, a and b can be expressed as:

$$a = \frac{d}{\sin \theta_{ae}} \tag{5.5}$$

$$b = f \sin \theta_{ae} \tag{5.6}$$

where

$$\theta_{ae} = \cot^{-1}\left(\frac{r_{\rm n} + \frac{f}{2}}{d}\right) \quad \text{where } d > r_{\rm n} \tag{5.7}$$

$$\theta_{ae} = \cot^{-1}\left(\frac{\sqrt{(2r_{\rm n} - d^2)} + \frac{f}{2}}{d}\right). \quad \text{where } d < r_{\rm n} \tag{5.8}$$

In Eqs. (5.7) and (5.8), d is depth of cut, f is feed, and $r_{\rm n}$ is nose radius of the cutting tool. Shear strength of the metal matrix k can be determined by implementing the Johnson–Cook model [40].

$$k = [A + B\varepsilon^n][1 + C \ln \dot{\varepsilon}]\left[1 - \left(\frac{T - T_r}{T_m - T_r}\right)^m\right] \tag{5.9}$$

where A, B, n, and m are the constants of Johnson–Cook's equation. Another parameter to be calculated is friction angle β. In any machining operation, a portion of energy is consumed to overcome friction between the chip and the tool's rake face. In the presence of hard and abrasive reinforcing particle when machining particle-reinforced MMCs, friction at the chip–tool interface is typically characterized by the two concepts, namely two-body rolling abrasion F_p and three-body rolling friction F_r [29, 30, 40]. Therefore, the total friction force is the combination of both factors as presented in Eq. (5.10).

$$F_f = F_p + F_r \tag{5.10}$$

For ease of calculations, the reinforcement particles can be assumed spherical [42], as presented in Fig. 5.7. Thus, F_p along the chip–tool interface can be expressed as:

$$F_p = N_p A_i 3\sigma_{y(\text{tool})} q \tag{5.11}$$

In Eq. (5.11), $\sigma_{y_{(\text{tool})}}$ is yield strength of tool material and q represents the fraction of the particles involved in two-body abrasion and its value depends on the cutting

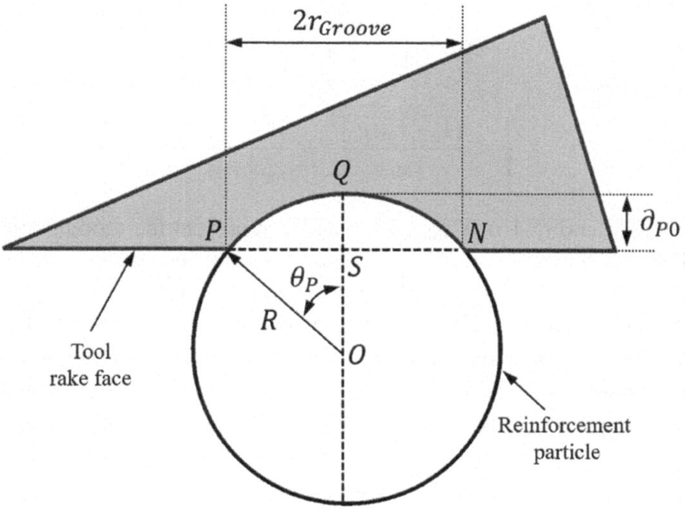

Fig. 5.7 Contact between abrasive particle and rake face of tool (with permission to reuse) [40, 44]

tool and workpiece materials. For instance, in machining alumina (Al_2O_3)-reinforced MMCs with ceramic tools, the value of q can be taken as 40% [43]. The number of abrasive particles involved in the tool–chip friction is indicated by N_p and can be calculated as follows:

$$N_p = \frac{V_p ba}{\pi R^2} \tag{5.12}$$

where V_p is the volume fraction of reinforcement particles, A_i is the contact area between a particle and the tool, R is the radius of particle, and a and b can be calculated using Eqs. (5.5) and (5.6). The contact area between a particle and the tool can be determined using Eq. (5.13) in which θ_p is the apex angle, as shown in Fig. 5.7, and is defined by Eq. (5.14).

$$A_i = \frac{R^2}{2}\left(\frac{\pi}{180}(2\theta_p) - \sin(2\theta_p)\right) \tag{5.13}$$

$$A_i = \frac{R^2}{2}\left(\frac{\pi}{180}(2\theta_p) - \sin(2\theta_p)\right) \tag{5.14}$$

In Eq. 5.14, ∂_{P_O} is the critical value of relative penetration and it can be calculated using Eq. (5.15).

$$\partial_{Po} = \left(\frac{9\pi}{4}\right)^2 \left(\frac{\sigma_{y(\text{tool})}}{\dot{Y}}\right)^2 R \tag{5.15}$$

$$\frac{1}{\dot{Y}} = \frac{(1-\nu_{\text{MMC}})^2}{Y_{\text{MMC}}} + \frac{(1-\nu_{\text{tool}})^2}{Y_{\text{Tool}}} \tag{5.16}$$

In Eqs. (5.15) and (5.16), Y_{MMC}, Y_{Tool} and $\dot{Y}$ represent the modulus of elasticity for the MMC workpiece material, tool, and chip–tool interface, respectively. The Poisson's ratio for the tool and workpiece material is expressed by ν_{tool} and ν_{MMC}. Total frictional force has another component yet to be determined that is the frictional force due to three-body rolling F_r. This parameter can be obtained using the following equation in which $\mu_{\text{three-body}}$ is the coefficient of friction for the three-body rolling and F_N is the total normal force at chip–tool interface.

$$F_r = \mu_{\text{three-body}} F_N \tag{5.17}$$

The coefficient of friction can be calculated based on models proposed by Venkatachalam and Liang [45] and Sin et al. [46] as presented in Eq. (5.18) where k_{tool} is the shear stress of the tool material and H_t is the tool hardness.

$$\mu_{\text{three-body}} = \frac{k_{\text{tool}}}{\pi H_t} \left(\frac{2R}{r_{\text{groove}}}\right)^2 \left\{ 1 - \left[1 - \left(\frac{r_{\text{groove}}}{2R}\right)^2\right]^{\frac{1}{2}} \right\} \tag{5.18}$$

$$r_{\text{groove}} = \sqrt{R^2 - (R - \delta_{Po})^2} \tag{5.19}$$

$$k_{(\text{tool})} = \frac{\sigma_{y(\text{tool})}}{2} = \frac{H_t}{6} \tag{5.20}$$

The other component that is required to determine the three-body rolling F_r in Eq. (5.17) is the total normal forces at chip–tool interface F_N. This parameter can be obtained using the following equation:

$$F_N = F_{N1} \times N_p \times q \tag{5.21}$$

where F_{N1} is the normal force on the individual abrasive particle and can be calculated using the approach proposed by Jiang et al. [42] as presented below:

$$F_{N1} = 2.9\pi R \sigma_{y(\text{tool})} \partial_{Po} \tag{5.22}$$

Calculating all of the required parameters using the previous equations, the friction angle β can be obtained.

$$\beta = \tan^{-1}\left(\frac{F_p + F_r}{F_N}\right) \tag{5.23}$$

Another parameter that affects the magnitude of forces during machining is the edge radius of the cutting tool. When the cutting edge is round, a portion of workpiece material will be squeezed underneath the cutting edge, which consequently introduces a new component of force called ploughing force. Based on the slip-line field model, the two components of ploughing force along and normal to the path of the tool can be calculated as follows [47]:

$$F_{\mathrm{CP}} = klr_{\mathrm{e}} \tan\left(\frac{\pi}{4} + \frac{\gamma}{2}\right) \tag{5.24}$$

$$F_{\mathrm{TP}} = klr_{\mathrm{e}}\left[1 + \frac{\pi}{2}\right] \tan\left(\frac{\pi}{4} + \frac{\gamma}{2}\right) \tag{5.25}$$

$$l = r_{\mathrm{n}}\left[\theta_{\mathrm{a}} + \sin^{-1}\left(\frac{b}{r_{\mathrm{n}}}\right)\right] + \frac{a - r_{\mathrm{n}}(1 - \cos\theta_{\mathrm{a}})}{\sin\theta_{\mathrm{a}}} \tag{5.26}$$

In Eqs. (5.24), (5.25), and (5.26), r_{e} is edge radius, r_{n} is nose radius, l is cutting width, γ is tool rake angle, and θ_{a} is approach angle. The last components of force to be calculated are those generated when the cutting tool fractures the abrasive reinforcement particles. The components of fracture force in cutting and trust directions can be expressed in terms of fracture energy denoted by U.

$$F_{\mathrm{CD}} = UabN_{\mathrm{p}} \times q \tag{5.27}$$

$$F_{\mathrm{TD}} = F_{\mathrm{cp}} \times \tan\delta \tag{5.28}$$

where δ shows the angle between the resultant force and cutting direction and can be expressed as follows:

$$\sin\delta = \frac{r_e(1 + \sin\gamma)}{2R + 2r_e} \tag{5.29}$$

Once all of the required components of force $(F_{\mathrm{CC}}, F_{\mathrm{TC}}, F_{\mathrm{CP}}, F_{\mathrm{TP}}, F_{\mathrm{CD}}, F_{\mathrm{TD}})$ are calculated, they can be substituted in Eqs. (5.1) and (5.2) to obtain the cutting force (F_{C}) and trust force (F_{T}).

5.6.2 *Machining of Fiber-Reinforced MMCs*

In comparison to particulate-reinforced MMCs, fiber-reinforced metal matrix composites are less likely to be machined using traditional machining operations. As a result, the research and studies about machining of fiber-reinforced MMCs, either in the form of fiber or whisker, are very limited. Turning and milling, as the two most widely used conventional machining operations, are not generally used for machining MMCs reinforced by continuous fibers. This is mainly due to the possibility of damages to the workpiece in terms of fiber pullout or fiber breakage. Because of the close relationship between the mechanical properties as well as the direction of fiber and mechanical characteristics of the whole bulk of composite material, any possible damage to the fibers extensively affects the mechanical properties of the entire part made of composite which is not desirable [20]. Another factor that limits to the application of traditional machining operations in machining of fiber-reinforced MMCs is the vulnerability of some types of fibers such as boron nitride and silicon carbide (SiC) to oxidation due to the heat generated during cutting.

Due to the limited availability of studies about machining fiber-reinforced MMCs, reviewing the research works about machining of fiber-reinforced polymer (FRP) composites is helpful to gain an initial understanding. Mechanics of chip formation during machining fiber-reinforced composite is greatly influenced by fiber orientation; while other parameters such as cutting conditions, tool geometry, matrix features, and fiber characteristics are not that influential [2]. The orientation of fiber (θ) can be classified as:

- $\theta = 0°$
- $0° \leq \theta \leq 90°$
- $90° \leq \theta \leq 180°$

5.6.2.1 Chip Formation When $\theta = 0°$

When the fiber orientation is parallel to the direction of cutting ($\theta = 0°$), the chip formation is mainly influenced by the cutting tool rake angle. In the case of rake angle equal to zero ($\alpha = 0°$), extreme compressive stresses are generated at the chip–tool interface (over the rake face of the cutting tool). As a result, the fibers are pushed forward in longitudinal directions that cause generating multiple cracks along the direction of fibers. Advancing further, tool pushes the fiber forward and propagates the initiated cracks. Chip slides over the rake face of the tool, and the process continues until the chip is completely fractured and the new chip is formed. This process is demonstrated in Fig. 5.8 [2, 48].

When the tool rake angle is positive ($\alpha \geq 0°$), since the tool engages the composite workpiece with its sharp cutting edge, instead of several randomly spaced cracks, a unique longitudinal crack is originated at the tool tip and propagates ahead of the tool. The material is separated along this crack, slides over the rake face, and eventually generates a cantilever-shaped chip [2]. This process continues until the stretched fibers are broken, and the chip is separated from the rest of material. Figure 5.9 shows the formation of the chip when machining fiber-reinforced composites under the ($\theta = 0°$) and ($\alpha \geq 0°$) conditions.

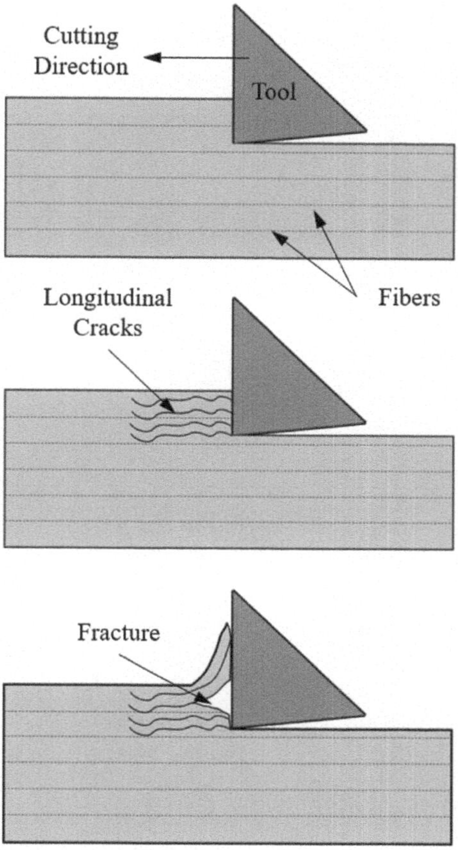

Fig. 5.8 Chip formation when fiber orientation $\theta = 0°$ and rake angle $\alpha = 0°$

Application of the negative rake angle ($\alpha \leq 0°$) is not frequent in the machining of composite materials; hence, not many experimental data are available in this regard.

5.6.2.2 Chip Formation When $0° \leq \theta \leq 90°$

When cutting fiber-reinforced MMCs with fiber orientation ($0° \leq \theta \leq 90°$), the cutting tool produces and localizes compressive stress while it engages the workpiece and advances through the workpiece body. This process is schematically shown through steps 1–4 in Fig. 5.10.

Because of this intense compressive stress, cracks are generated from fractured fibers, they propagate along the fiber direction toward the free surface of the composite body, and consequently the chip is formed. A detailed description of this phenomenon can be found in [48, 49].

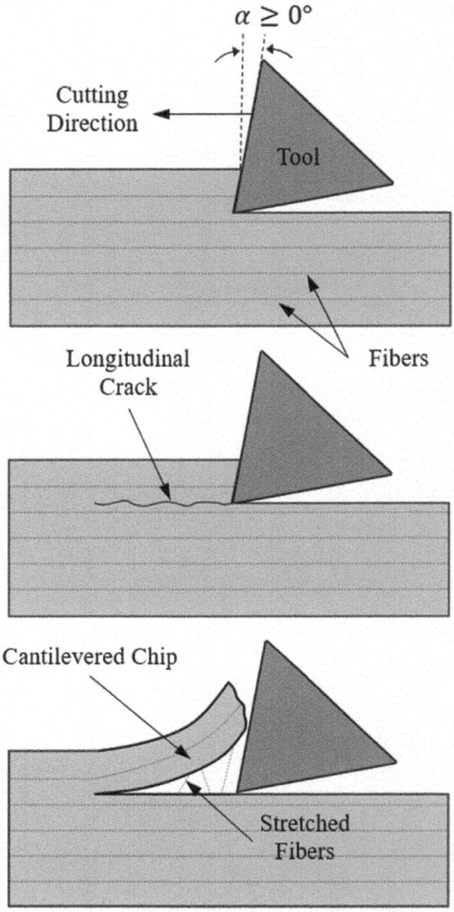

Fig. 5.9 Chip formation when fiber orientation $\theta = 0°$ and rake angle $\alpha \geq 0°$

5.6.2.3 Chip Formation When $90° \leq \theta \leq 180°$

Investigating the mechanics of chip formation when the fiber orientation ($90° \leq \theta \leq 180°$) is very complicated due to the nature of produced chips which are in the form of very tiny dusty particles [2].

5.7 Appropriate Tools Materials and Modes of Tool Wear

One of the greatest obstacles in widespread application of MMCs is their hard-to-cut nature, which makes them hard or sometimes very costly to produce with predefined level of dimensional accuracy and surface quality. Machining of MMCs is generally very expensive because it is very difficult to machine MMCs

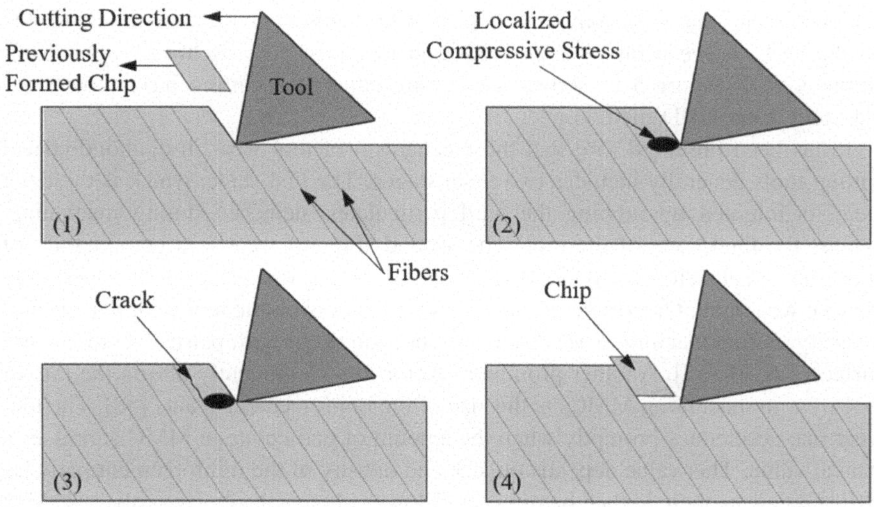

Fig. 5.10 Chip formation when fiber orientation $0° \leq \theta \leq 90°$

with the currently available cutting tools. Among the metal matrix composites, particulate-reinforced MMCs are very challenging materials when it comes to machining. Majority of reinforcement particles that are commonly used in reinforcing particulate MMCs are considerably harder than the conventionally utilized high-speed steel (HSS) and carbide tools [16]. As a result, a high rate of tool wear and deterioration can be observed when machining particulate MMCs using regular HSS and carbide tools. The followings are the candidate cutting tool materials for machining MMCs.

- Polycrystalline diamond (PCD)
- Cubic boron nitride (CBN)
- Polycrystalline cubic boron nitride (PCBN)
- Carbide tools with titanium nitride (TiN) coating
- Ceramic tools (Al_2O_3-TiC).

Despite their great performance in machining steels, carbide tools with titanium nitride (TiN) and titanium carbide (TiC) coatings demonstrate poor performance when they are used for machining MMCs [2, 50]. Hung [51] showed that when machining MMCs, these coatings are not able to resist when the reinforcements in the form of hard particles are rubbed against the tool; therefore, they provide insignificant improvement. The hard particles act like sandpaper and severely scratch the coating off the substrate. Great efforts have been made to increase the wear resistance of carbide tools by covering/protecting their surface by diamond coatings [7, 51–53]. Diamond coatings can be considered as the possible replacements for TiN and TiC coating in machining MMCs [50, 54]. It has been shown that if the coating strongly adheres to the tool substrate with the thickness around 500 μm, the wear resistance equal to that of polycrystalline diamond tools can be achieved using diamond-coated tools [55]. The diamond coating can be deposited on the tool surface using several deposition methods, e.g., chemical vapor deposition (CVD). Figure 5.11 shows the cutting edge of a cutting insert coated by diamond using CVD method.

It has been observed [56] that the mechanism of tool wear in diamond-coated cutting tools generally includes two main stages. The first stage, which is the flank wear, is initiated by rubbing the hard particulates, detached during machining, against the flank face of the tool. The second stage of wear is a combination of adhesion and abrasion when the MMC material, as the workpiece, starts adhering to the tool wear land. One of the primary affecting factors on the tool wear and surface integrity, during machining MMCs, is the abrasion of the hard particulate to the tool surface [16, 30, 57]. Another prominent factor that significantly affects the rate of tool wear in machining MMCs is the size of the reinforcement grains [58]. The tool wear rate accelerates promptly when the amount of particulate in MMC surpasses a critical value. This value depends on size and density of the reinforcement particles [30]. Owing to their higher hardness and lower chemical affinity, polycrystalline diamond (PCD) tools offer improved wear resistance and result in smoother machined surface when machining MMCs [16, 59, 60]. Figures 5.12 and 5.13

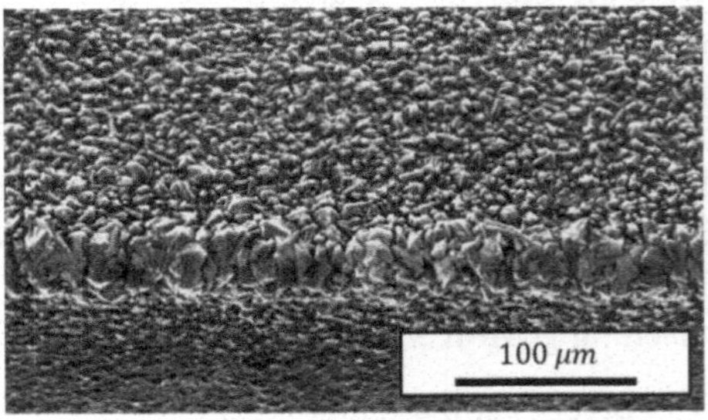

Fig. 5.11 Cutting edge of a CVD diamond-coated carbide insert (with permission to reuse) [54]

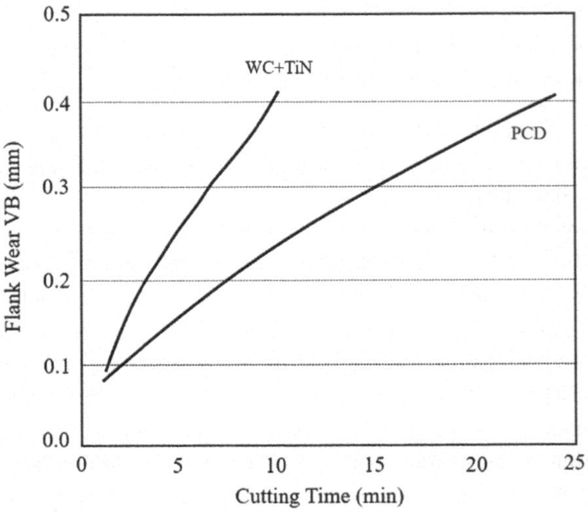

Fig. 5.12 Tool wear versus cutting time during turning of Al alloy +10% Al_2O_3. Cutting speed = 500 m/min, feed = 0.4 mm/rev, DOC = 1.5 mm (with permission to reuse) [7]

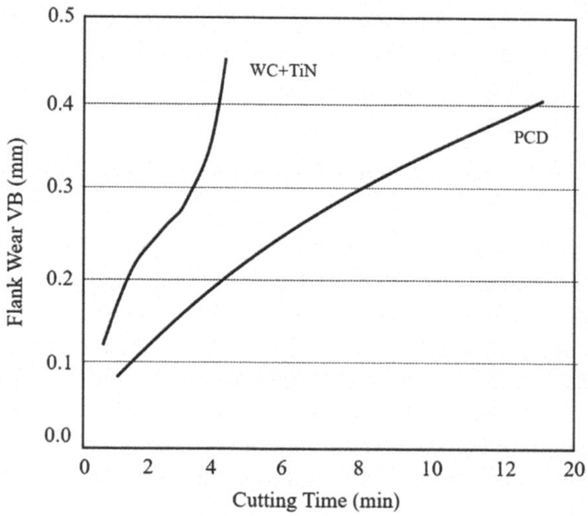

Fig. 5.13 Tool wear versus cutting time during turning of Al alloy +20% Al_2O_3. Cutting speed = 500 m/min, feed = 0.4 mm/rev, DOC = 1.5 mm (with permission to reuse) [7]

compare the flank wear of PCD tools with WC + TIN-coated cutting tools versus time during turning test [7].

Because of their good thermal resistance, hardness and thermal conductivity, polycrystalline cubic boron nitride (PCBN) tools are practical alternatives to the PCD tools in machining hard-to-cut materials. PCBN tools have better wear resistance in comparison to cemented carbides and ceramics in the machining of MMCs reinforced with SiC particles and also carbon-reinforced plastics [61, 62]. Hung [57] showed that in comparison to the WC tools, PCBN and PCD tools demonstrate better wear resistance in machining Al_2O_3 or SiC-reinforced MMCs. PCBN tools normally include grains of cubic boron nitride which are bounded together by means of metal binder or ceramic binders. Further improvement can be achieved by eliminating the metal or ceramics binder from PCBN tools, which results in binderless PCBN. Rahman et al. [63] and Neo et al. [62] determined that better surface finish and longer tool life (in terms of greater wear resistance) can be achieved in machining hard-to-cut materials such as titanium and stainless steels by replacing the conventional PCBN tools with the binderless ones [50].

Carbide tools exhibited a high rate of wear and poor surface finish in conventional turning of MMCs reinforced by Al/SiC particles [53, 58, 64]. Applying coolant when machining Al/SiC MMCs not only negatively affects the tool life by increasing the wear rate but also results in very poor surface finish. The hard SiC particles, which are used as reinforcements in Al/SiC MMCs, play the role of tiny cutting edges similar to a grinding wheel and deteriorate the cutting tool. When the cutting tool is deteriorated by abrasion, the machined surface will have very poor surface finish, which is not desirable. In some other cases, Al/SiC particles adhere to the tool because of the friction, high temperature, and pressure. While the cutting tool advances, extra particles join the already adhered ones and produce built-up edge ahead of the tool. Continuing this process generates very poor surface finish as the workpiece surface has been nibbled away instead of being cut. Therefore, achieving a good surface finish is always a challenging task when machining Al/SiC-reinforced MMCs in particular and MMCs in general [53].

5.7.1 Analytical Modeling of Wear Progression

As shown in Fig. 5.5 (see Sect. 5.6.1), examining the worn cutting tools when machining MMCs reveals the two-body and three-body abrasions. Experimental studies have also revealed that the progression of tool wear not only depends on the cutting parameters, but is also related to the tool geometry, particularly nose radius, as well as volume fraction and particle size of the reinforcement. After cutting time (t), the two-body abrasion can lead to a tool volume loss (V_{W2}) which can be presented, in the form of modified Archard's relation, as a function of cutting speed (V_c), normal load (N), and tool hardness (H_t) such that [32]:

$$\left(\frac{V_{W2}}{t}\right) \propto \left(\frac{N_{V_c}}{H_t}\right) \tag{5.30}$$

The rolling of the debonded hard particles between the cutting tool and matrix leads to three-body abrasion. As to a three-body abrasion model, particle size, volume fraction and hardness, and spacing between particles can be combined in an empirical form (modified Rabinowicz relation) as presented in Eq. 5.31:

$$\left(\frac{V_{W3}}{t}\right) \propto \left(\frac{NV_c}{xH_t}\left(\frac{H_t}{H_a}\right)^k\left(\frac{D}{d}\right)\left(\frac{f_V}{z(1-f_V)}\right)\right) \tag{5.31}$$

where V_{W3} represents the tool volume loss caused by three-body abrasion. x and k are empirical values known as Rabinowicz three-body abrasion wear constants [65]. The x and k are selected based on the tool and particle hardness ratio. H_a is the abrasive particle hardness, D is the particle diameter, d is the average spacing between the particles, f_V is the volume fraction of the particles, and z is the Weibull probability of undamaged particles [43]. Since same particle contributes into the two- and three-body abrasions, the total tool volume loss (V_w) can be given by [32]:

$$\frac{dV_W}{dt} = K\sqrt{\left(\frac{NV_c}{H_t}\right)^2 + \left(\frac{NV_c}{xH_t}\left(\frac{H_t}{H_a}\right)^k\left(\frac{D}{d}\right)\left(\frac{f_V}{z(1-f_V)}\right)\right)^2} \tag{5.32}$$

In Eq. (5.32), K is the proportionality constant, which can be calibrated for different tool/workpiece combinations and cutting conditions based on the method explained in [66]. Figure 5.14 shows a schematic representation of the volume loss of a tool used in bar turning operation. In Fig. 5.14, α is the clearance angle, γ is the rake angle, and R is the nose radius of the cutting tool.

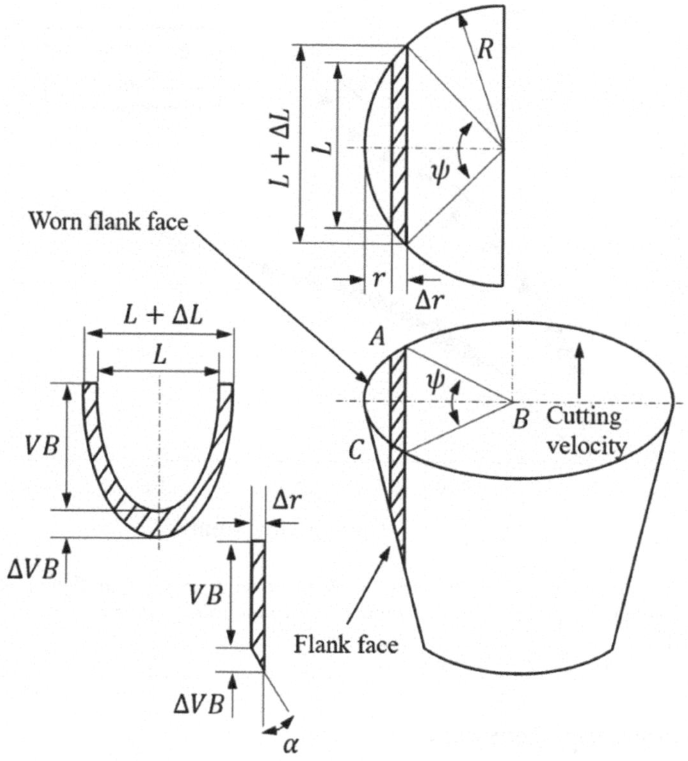

Fig. 5.14 Schematic geometry of the wear volume (with permission to reuse) [32]

In the case of turning operation with small depth of cut, it can be assumed that the nose radius of cutting tool is one order of magnitude larger than the depth of cut and feed. During machining, the length of flank wear (VB) increases with time, and thus based on Eq. (5.32) and geometrical relations extracted from Fig. 5.14, the flank wear rate is calculated as follows [32]:

$$\frac{dV_B}{dt} = \frac{3K}{4R\sin\left(\frac{\psi}{2}\right)\mathrm{VB}\tan\alpha}\left[\sqrt{\left(\frac{NV_c}{H_t}\right)^2 + \left(\frac{NV_c}{xH_t}\left(\frac{H_t}{H_a}\right)^k\left(\frac{D}{d}\right)\left(\frac{f_V}{z(1-f_V)}\right)\right)^2}\right]. \tag{5.32}$$

where ψ can be determined as a function of the depth of cut and tool nose radius. The predicted and measured progress of tool wear is shown in Fig. 5.15 under different volume fractions.

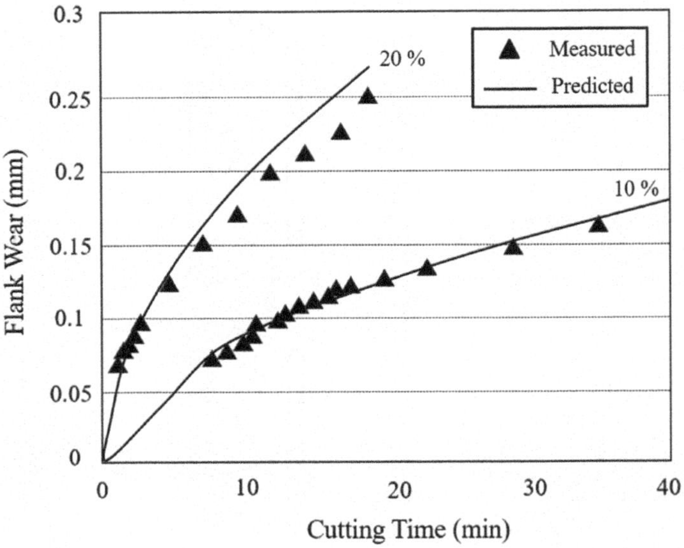

Fig. 5.15 Wear progression of an uncoated tungsten carbide insert when machining 6061 MMC with different volume fractions ($V_c = 12$ m/min, $f = 0.1$ mm, DOC $= 0.15$ mm) (with permission to reuse) [32]

5.8 Concluding Remarks

Metal matrix composites are classified among the hard-to-cut materials due to the presence of hard reinforcement with abrasive nature in their structure. These abrasive particles cause some difficulties in the machining of these materials and consequently impose a huge barrier in their widespread application. Several studies on machining of MMCs illustrate that PCD tools are the best option for machining MMCs. PCD tools are harder than the typical particulate reinforcements such as Al_2O_3 and SiC, and they do not chemically react to workpiece materials; thus, they provide an acceptable tool life when machining Al/SiC MMCs. In addition to PCD, other cutting tool materials such as chemical vapor deposition (CVD) inserts, TiN-, Ti(CN)-, and Al_2O_3-coated tools can also be applied in the machining of MMCs. Among CVD inserts, TiN-, Ti(CN)-, and Al_2O_3-coated tools, CVD inserts show improved performance capabilities in comparison to the other tool materials. Generally, in comparison to the Al_2O_3/TiC- and TiN-coated carbide tools, PCD tools exhibited the smallest tool wear due to their high thermal conductivity and low coefficient of friction.

References

1. Chawla KK. Composite materials: science and engineering. Springer Science & Business Media (2012)
2. Hocheng H. Machining technology for composite materials. In: Hocheng H, editor. Cambridge, UK: Woodhead Publishing; 2012.
3. Kelly A, Zweben CH. Comprehensive composite materials. Amsterdam: Elsevier; 2000.
4. Chawla N, Chawla KK. Metal matrix composites. New York: Springer; 2013.
5. Mortensen A, Llorca J. Metal matrix composites. Annu Rev Mater Res. 2010;40:243–70.
6. Manna A, Bhattacharyya B. Investigation for optimal parametric combination for achieving better surface finish during turning of Al/Sic-Mmc. Int J Adv Manuf Technol. 2004;23(9–10):658–65.
7. Durante S, Rutelli G, Rabezzana F. Aluminum-based MMC machining with diamond-coated cutting tools. Surf Coat Technol. 1997;94:632–40.
8. Chawla KK, Chawla N. Metal-matrix composites. Wiley Online Library; 2004.
9. Cyriac AJ. Metal matrix composites: history, status, factors and future. Oklahoma State University; 2011.
10. Dow NF. Study of stresses near a discontinuity in a filament-reinforced composite metal. DTIC Document; 1963.
11. Wadsworth J, Lesuer DR. Ancient and modern laminated composites—from the Great Pyramid of Gizeh to Y2k. Mater Charact. 2000;45(4):289–313.
12. Jech R, McDanels D, Weeton JW. Metals reinforced with fibers. MET PROG. 1960;78 (6):118–21.
13. Peters ST. Handbook of composites. US: Springer; 2013.
14. Kilickap E, Inan A. A study on machinability of Al Si7 Mg2/SiCp metal matrix composite. Int J Mach Mach Mater. 2006;1(4):463–75.
15. Chawla N, Chawla KK. Metal matrix composites. Berlin: Springer; 2006.
16. El-Gallab M, Sklad M. Machining of Al/Sic particulate metal matrix composites: Part II: workpiece surface integrity. J Mater Process Technol. 1998;83(1):277–85.
17. Monaghan J. The use of a quick-stop test to study the chip formation of a Sic/Al metal matrix composite material and its matrix alloy. Int J Fatigue. 1996;3(18):213.
18. Clyne T, Withers P. An introduction to metal matrix composites. Cambridge University Press; 1995.
19. Mallick PK. Fiber-reinforced composites: materials, manufacturing, and design. CRC press; 2007.
20. Davim JP. Machining of metal matrix composites. Springer; 2012.
21. Rawal SP. Metal-matrix composites for space applications. JOM. 2001;53(4):14–7.
22. Natarajan N, Vijayarangan S, Rajendran I. Wear behaviour of A356/25sic P aluminium matrix composites sliding against automobile friction material. Wear. 2006;261(7):812–22.
23. Suresh, S. fundamentals of metal-matrix composites. Elsevier; 2013.
24. Deuis R, Subramanian C, Yellup J. Dry sliding wear of aluminium composites—a review. Compos Sci Technol. 1997;57(4):415–35.
25. Chawla N, Chawla K. Metal-matrix composites in ground transportation. JOM. 2006;58 (11):67–70.
26. Teti R. Machining of composite materials. CIRP Ann Manuf Technol. 2002;51(2):611–34.
27. Dandekar CR, Shin YC. Modeling of machining of composite materials: a review. Int J Mach Tools Manuf. 2012;57:102–21.
28. Pramanik A, Zhang L, Arsecularatne J. An FEM investigation into the behavior of metal matrix composites: tool-particle interaction during orthogonal cutting. Int J Mach Tools Manuf. 2007;47(10):1497–506.
29. Sahin Y, Sur G. The effect of Al_2O_3, Tin and Ti (C, N) based CVD coatings on tool wear in machining metal matrix composites. Surf Coat Technol. 2004;179(2):349–55.

30. Li X, Seah W. Tool wear acceleration in relation to workpiece reinforcement percentage in cutting of metal matrix composites. Wear. 2001;247(2):161–71.
31. Ciftci I, Turker M, Seker U. Evaluation of tool wear when machining Sic P-reinforced Al-2014 alloy matrix composites. Mater Des. 2004;25(3):251–5.
32. Kishawy H, Kannan S, Balazinski M. Analytical modeling of tool wear progression during turning particulate reinforced metal matrix composites. CIRP Ann Manuf Technol. 2005;54 (1):55–8.
33. Stephenson DA, Agapiou JS. Metal cutting theory and practice, vol 68. CRC press; 2006.
34. Shaw MC, Cookson J. Metal cutting principles, vol. 2. New York: Oxford University Press; 2005.
35. Trent EM. Metal cutting. Butterworth-Heinemann Ltd.; 1991. p. 273.
36. Pramanik A, Zhang L, Arsecularatne J. Machining of metal matrix composites: effect of ceramic particles on residual stress, surface roughness and chip formation. Int J Mach Tools Manuf. 2008;48(15):1613–25.
37. Ng E-G, Aspinwall DK. The effect of workpiece hardness and cutting speed on the machinability of Aisi H13 hot work die steel when using PCBN tooling. J Manuf Sci Eng. 2002;124(3):588–94.
38. Karthikeyan R, Ganesan G, Nagarazan R, Pai B. A critical study on machining of Al/Sic composites. Mater Manuf Processes. 2001;16(1):47–60.
39. Lin J, Bhattacharyya D, Lane C. Machinability of a silicon carbide reinforced aluminium metal matrix composite. Wear. 1995;181:883–8.
40. Sikder S, Kishawy H. Analytical model for force prediction when machining metal matrix composite. Int J Mech Sci. 2012;59(1):95–103.
41. Colwell LV. Predicting the angle of chip flow for single-point cutting tools. Trans ASME. 1954;76:199.
42. Jiang J, Sheng F, Ren F. Modelling of two-body abrasive wear under multiple contact conditions. Wear. 1998;217(1):35–45.
43. Kishawy H, Kannan S, Balazinski M. An energy based analytical force model for orthogonal cutting of metal matrix composites. CIRP Ann Manuf Technol. 2004;53(1):91–4.
44. Dabade UA, Dapkekar D, Joshi SS. Modeling of chip-tool interface friction to predict cutting forces in machining of Al/SICp composites. Int J Mach Tools Manuf. 2009;49(9):690–700.
45. Venkatachalam S, Liang SY. Effects of ploughing forces and friction coefficient in microscale machining. J Manuf Sci Eng. 2007;129(2):274–80.
46. Sin H, Saka N, Suh N. Abrasive wear mechanisms and the grit size effect. Wear. 1979;55 (1):163–90.
47. Waldorf DJ. A simplified model for ploughing forces in turning. J Manuf Process. 2006;8 (2):76–82.
48. Lopresto V, Santo L, Caprino G. Mechanisms of chip generation in orthogonal machining of unidirectional fiber reinforced plastics.
49. Pwu H, Hocheng H. Chip formation model of cutting fiber-reinforced plastics perpendicular to fiber axis. J Manuf Sci Eng. 1998;120(1):192–6.
50. Ding X, Liew W, Liu X. Evaluation of machining performance of MMC with PCBN and PCD tools. Wear. 2005;259(7):1225–34.
51. Hung N, Boey F, Khor K, Oh C, Lee H. Machinability of cast and powder-formed aluminum alloys reinforced with Sic particles. J Mater Process Technol. 1995;48(1–4):291–7.
52. Iuliano L, Settineri L, Gatto A. High-speed turning experiments on metal matrix composites. Compos A Appl Sci Manuf. 1998;29(12):1501–9.
53. Looney L, Monaghan J, O'Reilly P, Taplin D. The turning of an Al/Sic metal-matrix composite. J Mater Process Technol. 1992;33(4):453–68.
54. Chou YK, Liu J. CVD diamond tool performance in metal matrix composite machining. Surf Coat Technol. 2005;200(5):1872–8.
55. D'Errico G, Calzavarini R. Turning of metal matrix composites. J Mater Process Technol. 2001;119(1):257–60.

56. Andrewes CJ, Feng H-Y, Lau W. Machining of an aluminum/Sic composite using diamond inserts. J Mater Process Technol. 2000;102(1):25–9.
57. Hung N, Loh N, Xu Z. Cumulative tool wear in machining metal matrix composites part II: machinability. J Mater Process Technol. 1996;58(1):114–20.
58. Weinert K, König W. A consideration of tool wear mechanism when machining metal matrix composites (MMC). CIRP Ann Manuf Technol. 1993;42(1):95–8.
59. Hooper R, Henshall J, Klopfer A. The wear of polycrystalline diamond tools used in the cutting of metal matrix composites. Int J Refract Metal Hard Mater. 1999;17(1):103–9.
60. Chambers A. The machinability of light alloy MMCs. Compos A Appl Sci Manuf. 1996;27 (2):143–7.
61. Ferreira J, Coppini N, Miranda G. Machining optimisation in carbon fibre reinforced composite materials. J Mater Process Technol. 1999;92:135–40.
62. Neo K, Rahman M, Li X, Khoo H, Sawa M, Maeda Y. Performance evaluation of pure CBN tools for machining of steel. J Mater Process Technol. 2003;140(1):326–31.
63. Rahman M, San Wong Y, Zareena AR. Machinability of titanium alloys. JSME Int J Ser C Mech Syst Mach Elem Manuf. 2003; 46(1):107–15.
64. Yanming Q, Zehua Z. Tool wear and its mechanism for cutting sic particle-reinforced aluminium matrix composites. J Mater Process Technol. 2000;100(1):194–9.
65. Huang Y, Liang SY. Modeling of CBN tool flank wear progression in finish hard turning. Trans. Am. Soc. Mech. Eng. J. Manuf. Sci. Eng. 2004;126(1):98–106.
66. Zhang Z, Zhang L, Mai Y-W. The running-in wear of a steel/SICp-Al composite system. Wear. 1996;194(1–2):38–43.

Chapter 6
Ceramics

Abstract Ceramics are quite varied in terms of industrial applications while they are unique in terms of mechanical properties and manufacturing methods. Ceramic materials and their unique characteristics are one of the main fields in which a great rate of research progress and development have been done in the past, yet a greater progress can be foreseen in future. Due to their exceptional physical properties, components made from ceramic need to be produced using advanced and well-controlled manufacturing processes. The present chapter covers topics such as description of ceramic materials, history of evolution, and their current applications in industry. It also discusses the machining and machinability of ceramic materials. This chapter can be divided into five main sections: introduction, application in industry, manufacturing process, challenges, and a brief overview of non-traditional machining of ceramics. Relevant background information about ceramics as well as their applications in different industrial sectors such as automotive, aerospace, medical, military, textile, and more is also reviewed in this chapter. The current chapter also presents the current challenges in machining ceramics, applicable tools and manufacturing processes, and other influential contributing factors to the machining and machinability of ceramics.

6.1 Introduction

The word ceramic is originally rendered from the Greek word *keramikos* [1] that stands for pottery. *Keramikos* itself is derived from *keramos* which has a meaning close to burnt stuff [2]. Due to their special characteristics, finding a unique definition to describe ceramics, as a category of materials, is a challenging task. In the traditional texts, ceramics are defined as "of or having to do with pottery" [2, 3]. This definition directly refers to the ancient era when early humankind excavated earthy cay, mixed the cay with water to form clay, created an outline and dried the object either in the sun or by means of fire. These hard and brittle materials can be considered as the early ceramics made by our ancestors. However, the above-mentioned definition does not accurately describe the new generation of synthetic ceramics, which were initially introduced in the twentieth century. A new

H. A. Kishawy and A. Hosseini, *Machining Difficult-to-Cut Materials*, Materials Forming, Machining and Tribology, https://doi.org/10.1007/978-3-319-95966-5_6

definition for ceramic materials was proposed by Kingery [4], which described ceramics as "solid articles which are composed mostly of inorganic, non-metallic materials as their essential components." Based on this definition, ceramics is one of the three primary groups in which solid materials are categorized, namely ceramics, organics, and metals [2]. Therefore, any material that is not a metal or an organic material can be considered as ceramic.

Another definition for ceramics was proposed by Carter and Norton [5] where ceramics are defined as "mixed bonding or a combination of covalent, ionic, and sometimes metallic materials" [5]. Their definition describes ceramics as an array of interconnected atoms where no distinct molecules can be observed. This definition separates ceramics from molecular solids (e.g., iodine crystals and paraffin wax). Generally speaking, ceramics are combinations of metals or metalloids and non-metals, and they are typically found in the form of oxides, nitrides, and carbides. Ceramics can be either crystalline or non-crystalline and generally exhibit high melting points, low conductivity, and high strength. Non-crystalline ceramics, also known as glass, can be formed by melting and casting, while crystalline ceramics are not easy to form and cannot go through many processing techniques.

Ceramics can also be categorized into advanced or traditional ones. Traditional ceramics consist of cement, silicate glass, and clay products, whereas advanced ceramics include non-silicate glasses, carbides (e.g., SiC), pure oxides (e.g., Al_2O_3), and nitrides (Si_3N_4). Among these two categories, the former are mainly used in basic objects due to their simplicity while the latter are often used for more advanced applications such as electronic. Ceramics are promising materials due to their strength, hardness, and heat-resistant properties. Their applications cover a wide range from tableware to space shuttles. Large-scale applications such as buildings and cross-country power lines, to miniaturized applications on an atomic level, such as superconductors are the other examples of ceramics' use in today's life.

6.2 Historical Background and Evolution of Ceramics

Since as early as 29,000 B.C., ceramics have been used by ancient human civilizations [6]. The very first examples of ceramics can be attributed to the creation and use of pottery, which is one of the oldest technologies and human invention, dated back to 29,000–25,000 B.C. The origin of ceramics can be attributed to when humans first discovered that clay can be used to make objects by mixing water and then heating it up.

However, it was not until roughly 14,000 B.C. when the first signs of ceramic pottery began to appear [7]. Pottery is made through the process of using ceramic (clay in most instances) and forming it into the desired shape [8]. The object is then heated in high temperatures in the range of 1000 up to 1400 °C to evaporate the existing water. This process hardens the object and increases its strength. There are several different types of clay, and the properties of each type determine how the pottery objects react upon the evaporation of the water when heated. The first use of pottery objects was to store food and water that eventually led to the use and make of clay bricks or tiles [5]. The Jomon people of Japan, who were a society of hunter–gatherers, learned to create functional containers by kneading, forming, and firing clay. In the early use of ceramics, clay was formed into shapes, then fired on a pile of wood in a method called open-firing. As history progressed, new firing methods that used tunneled, sloping kilns were developed, which allowed the clay to be fired in a tunnel at higher temperatures, for a longer period of time [9]. This method allowed for cooling before exiting the tunnel, which was more efficient due to the fact that the heat released from the materials in the latter part of the tunnel could be absorbed and recycled. Moreover, since these kilns were larger and could be used for longer periods of time, ceramic technology began to improve as humans were able to produce ceramic vessels in larger quantities.

The beginning of the manufacture of glass goes hand in hand with the manufacture of ceramics. The first known production of glass was in 8000 B.C., by ancient Egyptian civilizations [6]. However, it was not until much later that glass was used as a separate material. The overheating of pottery in kilns produced a colored glaze on the vessels, which allowed them to stay water resistant. However, glass itself was not manufactured separately for its own purpose until 1500 B.C.

Large-scale manufacturing of ceramics began in the middle ages when the metal industry introduced furnaces that could be used for metal forging and casting [6]. These high-volume high-temperature furnaces were able to efficiently produce ceramic vessels at a very large scale. In the 1720s, Europeans began to master the art of manufacturing fine, translucent porcelain for aesthetic purposes such as dental prosthetics. Usually, feldspar was used as a substitute for lime, and the mixture was placed under very high firing temperatures, which contribute to its smooth, finer quality porcelain. The first porcelain dentures were manufactured in 1774 replacing elephant and hippopotamus ivory as the material of choice for dentures. By 1808, porcelain teeth were becoming widely used and embedded with platinum pins, which provided mechanical versatility to the teeth.

In the second half of the nineteenth century, ceramic materials began to be used for purposes other than for daily life (e.g., food and water storage) applications. With the development of electricity and the invention of the electric lightbulb, the insulating properties of ceramic materials made them ideal for use in new technologies such as cars, radios, televisions, and computers. In addition, the large-scale manufacture of ceramic materials allows large-scale applications such as power lines, to provide electricity to homes. At the time, other insulators such as paper and wood were also being used; however, the resistance of ceramics to high temperatures and humidity compared to their paper and wood counterparts made them more reliable as an insulator in electrical applications [7].

In 1911, Dutch physicist and Nobel laureate Heike Kamerlingh Onnes cooled mercury down to a temperature of −269 °C and discovered that it conducted electricity without losing any energy. This phenomenon that only occurred at extremely low temperatures was called superconductivity. However, the first major breakthrough in the field of superconductivity came from a ceramic, called yttrium barium copper oxide. That ceramic showed electrical conductivity at normal temperatures and superconductivity once cooled down to about −193 °C. Since this discovery, scientists have begun to work on creating new ceramics at atomic levels, with the ultimate goal of creating superconductors that can function at practical temperatures.

With the invention of the transistor in the twentieth century, ceramics began to become commonplace in consumer products. Early computers that required large vacuum tubes relied heavily on the insulating properties of ceramic materials. In fact, one of the first capacitors invented, the Leyden jar, was a ceramic capacitor. Before the World War II, mica was the most commonly used dielectric for capacitors [10]. The beginning of World War II, however, brought about ceramic packages that could insulate electric components from external moisture and light while still maintaining the electrical performance [7]. This was a crucial development as early transistors and electric circuits were very sensitive to external environmental conditions.

As the development of electronics continued, new types of combined ceramics were able to hold dielectric and magnetic properties, allowing for the transmission of electric force without conduction [7]. At the end of World War II, barium titanate ceramics were discovered and initially implemented as underwater transducers, within communication devices and for dielectric components such as capacitors [11]. Because of the lack of communication between countries during World War II, barium titanate ceramics were discovered on separate occasions by different researchers, independently. It was first discovered in 1942 by American researchers E. Wainer and N. Salomon; then again in Japan in 1944 by T. Ogawa, and in the Soviet Union by B. M. Vul in the same year [11]. Initially, researchers believed these ceramics were typical ferroelectric materials until their piezoelectric properties were discovered by S. Roberts in 1947 [11]. By the early 1950s, piezoelectric transducers that were based on barium-titrate ceramics were becoming widely used in consumer and military applications. However, in 1954, a new type of ceramic, lead-zirconate-titanate, was discovered, and replaced barium titanate in many piezoelectric applications [11]. One of the most important properties of lead-zirconate-titanate ceramics and the reason why they are popular is their high electromechanical coupling factor, and good frequency–temperature characteristics. Moreover, they are not soluble in water and are easy to produce [11]. The progress of ceramic materials allowed for a continuous reduction in the size of capacitors and inductors in electronics, which in turn gave way to the manufacture of even smaller electronic devices and miniaturized electrical components. With the invention of the multilayer ceramic capacitor in the 1970s, ceramic capacitors were now able to hold even higher voltages depending on whether the individual ceramic layers were arranged in parallel or series [10].

6.3 Material Structure of Ceramics

As mentioned earlier, ceramics are made up of oxides, carbides, nitrides, and silicides. Sometimes borides, phosphides, and selenides are used for certain applications [12]. These non-metal substances are combined with some metals or metalloids such as aluminum, zinc, titanium, or silicon to produce the structure of ceramics. They exhibit especial characteristics that no other type of material offers including low densities, high heat capacity, low coefficient of thermal expansion, high melting point, and consequently resistance to high temperature, corrosion resistance, good electrical insulation, and ultimately high strength and hardness.

Ceramics owe their superior characteristics to the ionic bonds that tie their atoms together. These bonds are the strongest bonds possible when compounds are formed. Due to the strong ionic bonds between the atoms of ceramic materials, a high magnitude of force and energy is required to separate these bonds. As a result, ceramics possess very high strength values. Ceramics also have high wear resistance, which means that their surface cannot be easily deteriorated or scratched. Ionic bonds also allow the structure to stay stable when any external disruptions such as heat, chemicals, or electricity occur. The electrons from both substances in the compound are given or lost to one another, which allow each substance to gain a full valence shell. This means that there are no free electrons within the compounds, which means that electricity and heat will not be conducted through the compound. In addition, the ionic bonds in ceramics allow the compounds to resist any acids applied to them. Acids are made up of hydrogen atoms, which are mainly positive, and they want to bond with any loose electrons they meet. Since there are no free electrons in ceramics, acids will not have any effect on the compound structure. The ionic bonds in ceramics also allow for lower density values compared to metals. This offers more freedom in design including up to 40% lower component weight, an increase in the rigidity and strength of the material, and eventually a significant reduction in cost in comparison to higher density compounds. On the other hand, due to their low heat conductivity and low coefficient of thermal expansion, the rate at which ceramics deform with increase in temperature is much lower than a material with a high thermal expansion coefficient. These two features combined together make ceramic materials an extremely good heat insulator capable of storing large amounts of heat with very little change in dimensions and overall shape.

Despite their desired characteristics, ionic bonds have some inherent features that are considered undesirable. Due to the strength of ionic bonds and lack of space for atoms to move, ceramics have almost no freedom to deform into certain shapes. As a result, ceramics are very brittle and susceptible to fracture. In cases when enough force is applied to break an ionic bond, the entire microstructure loses stability and fractures entirely along a certain plane. Ceramics are also vulnerable to shear forces, which cause them to crack and/or break apart without significant deformation. These undesirable features introduce several difficulties when it comes to forming ceramics into complex shapes and joining ceramic parts together or to the other types of materials. Hence, when considering ceramics as the material of choice for a certain application, extra care and attention must be paid to their mechanical properties to ensure the possibility of manufacture with least amount of issues.

Although achieving a comprehensive categorization for ceramics is not easy, the following materials are considered to be ceramic by most of the materials engineers [2].

6.3.1 Polycrystalline Ceramics Made by Sintering

This category includes those ceramics that can be made by mixing powder with water or any other types of polymers or solvent and forming them to the desired shape by pressing them against each other. The resultant object is then sintered at high temperature to let the particles bond together and form a polycrystalline ceramic [2].

6.3.2 Glass

Despite polycrystalline ceramics where the mixture of raw material is sintered at high temperature to form a solid object, in this category of ceramic materials, the mixture of raw materials is heated high enough to obtain a homogeneous liquid. The liquid is then shaped and cooled in such a way to prevent forming any grains or crystals [2].

6.3.3 Glass Ceramics

Glass ceramics differ from glass in their crystalline structure. As it was previously described, glasses are formed by cooling the liquid in a specific manner to prevent forming crystals while in glass ceramics crystal structure is preferred. In this category of ceramic materials, a nucleating agent is added and the glass is formed. The resultant glass is then heat-treated at a certain temperature below the melting point to initiate crystallization. In this case, the atoms in the glass reorganize around the nucleating agent atoms and generate polycrystalline structure called glass ceramics [2].

6.3.4 Single Crystals of Ceramic Compositions

Some single crystal ceramics are produced by heating ceramic powders while some of them are generated at room temperature by growing crystals from supersaturated solution. The characteristics of single crystals can be different from the polycrystalline forms of the same mineral. The single crystal form of natural minerals like quartz, ruby, and diamond is rarely found in nature; however, they can be made synthetically in laboratories [2].

6.3.5 Chemical Synthesis or Bonding

This category of ceramic materials is formed at room temperature using chemical reactions. Good examples of chemical synthesis or bonding ceramics are portland cement and plaster [2].

6.3.6 Natural Ceramics

Those ceramics that can be found in nature belong to this category of ceramic materials. Quartz crystal with wide application in piezoelectric applications is a good example of this category of ceramics [2].

6.4 Characteristics of Ceramic Materials

6.4.1 Brittleness

Ceramic materials are generally very brittle due to the special type of connection between the atoms, which is called mixed ionic–covalent bonding. It must be mentioned that ceramics may not show brittleness at high temperature. Thus, it would be more precise to say that the majority of ceramics show brittle characteristics at room temperature while such a behavior may not be observed at elevated temperature. Due to their brittle nature, ceramics normally have low toughness; however, their toughness can be increased by producing ceramic composites [2, 5].

6.4.2 Poor Electrical and Thermal Conductivity

As mentioned earlier, due to mixed ionic–covalent bonding, the valence electrons, which their movement is a directly responsible for electrical and thermal conductivity, are not free to move in ceramic materials. This results in poor electrical and thermal conductivity of most of the ceramics, although some exceptions may also exist [5].

6.4.3 Compressive Strength

An engineering component experiences different types of loading during its service life; therefore, its behavior under these situations must be clearly determined prior to design process starts. Ceramic materials are more vulnerable in tension than in compression. This characteristic can be very critical when a ceramics component is to be used in load-bearing applications. To prevent failure, it must be ensured that the stress distribution within the ceramic component is compressive [2, 5].

6.4.4 Chemical Insensitivity

Ceramics are not chemically reactive materials and the majority of them can sustain in severe chemical and thermal environments. Pyrex glass is a good example of ceramic materials, which is used in a broad range of applications from bakeware to chemistry laboratories' equipment. Pyrex glass shows a great stability at high temperatures (up to 1100 °K), and it also resists in many corrosive chemical environments [5].

6.5 Industrial Applications of Ceramics

Use of ceramics has come a long way since its first discovery. Traditionally, ceramics were used for structural materials such like cement and bricks, earthenware, white wares, plasters, and glasses. White wares are the ceramics that can be found in spark plugs, in laboratory equipment, and in high-class potteries. Application of ceramics in construction is a multi-billion-dollar industry. The amount of glass that is produced in a year is estimated to be enough to cover a 200-foot-wide glass highway from New York to Los Angeles [13]. Application of ceramics has significantly increased through the last decades, and it is no longer limited to the daily traditional usages. Throughout the years, ceramics have been modified to satisfy the needs of twenty-first century by various methodologies and techniques. Nowadays, ceramics are used in automotive, aerospace, medical, military, power generation, textile, and leisure industries. Applications of ceramic materials in modern industry can be categorized as follows [5]:

6.5.1 Structural Applications

Ceramic materials can be used in several structural applications such as heat exchangers and engine parts. They can also be used as wear-resistant components and cutting tools [5]. Such applications require special physical (mechanical) and chemical characteristics to survive in harsh working conditions. The ceramic materials to be used in such applications must be chemically inert and creep resistant as well. They should also have low density, high hardness and maintain their hardness and strength at elevated temperatures. Appropriate ceramics for structural applications include silicon nitride (Si_3N_4), silicon carbide (SiC), zirconia (ZrO_2), boron carbide (B_4C), and alumina (Al_2O_3) [5].

6.5.2 Electronic Applications

Ceramic materials are also used in electric and electronic industries for applications such as heat insulators, capacitors, dielectrics, micro-electro-mechanical systems (MEMS), substrates, and packages for integrated circuits [5]. Ceramics exist in several electrical devices such as microwaves, ovens, and even in technologies such as phones and computers. Ceramics such as barium titanate ($BaTiO_3$), zinc oxide (ZnO), lead-zirconate-titanate ($Pb(Zr_xTi_{1-x})O_3$), and aluminum nitride (AlN) are the potential candidates for these types of applications [5].

6.5.3 Bio-Applications

Medical treatments are always progressing and evolving to give patients a better lifestyle. Material selection in bio-industry is a very crucial step since the material would directly interact with the human body. The choice of ceramics for medical and bio-application is a modern development owing to their chemical inertness. One of the primary applications of ceramics as biomaterials is in implants where the component must not react with the human's body [5, 14, 15]. In surgical implants, the main job of ceramic is to replace the damaged tissue and be compatible to the physiological environment. Ceramics are most commonly used for hip transplants, knee prostheses, ankle joints, elbow, shoulders or fingers, as it provides strength, reliability, durability, and toughness and with the combination of bio-inert for a seamless replacement. These ceramics can be in the form of very durable metal oxides such as alumina (Al_2O_3) and zirconia (ZrO_2) which are among the appropriate ceramics for bio-applications [5]. Some bone implants are made from ceramics due to the material's toughness and bio-inert properties. Most of the hip implants use alumina ball for the femoral-head component. The field of dentistry uses ceramics for tooth implants and braces. Ceramic braces are transparent allowing for a more appealing choice. They are used as replacements of traditional metal braces, caps, or dental implants, which were very unappealing.

6.5.4 Coating Applications

One of the important applications of ceramic materials is coating which is normally performed for economic reasons. Coatings are normally applied when special surface characteristics must be achieved while the properties of the substrate are not as important. In such cases instead of making a part entirely from an expensive material, it can be made from a cheaper material and then coated with the desired material, which is normally expensive, to obtain the required surface characteristics [5]. A good example of using ceramics as a coating would be covering the cutting inserts with diamond coating to improve hardness and wear characteristics.

6.5.5 Composites Applications

Composite materials are normally produced by combining two different phases, namely matrix and fiber (reinforcement) to utilize the preferred characteristics of both parties while reducing their weakness. Ceramic materials can be used either as matrix phases in ceramic matrix composites (CMC) or as reinforcement in metal matrix composites (MMC). Ceramics are very brittle; hence, their fracture toughness can be significantly improved by using them as a matrix phase and adding whiskers or fibers of different materials as reinforcement. In MMCs, ceramics are generally used as reinforcement to enhance the wear and creep resistance of metal [5]. Table 6.1 shows the current and future applications of advanced ceramics in different aspects of industry as well as human life.

Table 6.1 Typical application of ceramics in different aspects of industry (with permission to reuse) [16]

Mechanical Engineering	**Aerospace**	**Automotive**
Cutting tools and dies	Fuel systems and valves	Heat engines
Abrasives	Power units	Catalytic converters
Precise instrument parts	Low-weight components	Drivetrain components
Molten metal filter	Fuel cells	Turbines
Turbine engine components	Thermal protection systems	Fixed boundary recuperators
Low-weight components for rotary equipment	Turbine engine components	Fuel injection components
Wearing parts	Combustors	Turbocharger rotors
Bearings	Bearings	Low heat rejection diesels
Seals	Seals	Water-pump seals
Solid lubricants	Structures	

(continued)

Table 6.1 (continued)

Defense industry	Biological and chemical processing engineering	Electrical, magnetic engineering
Tank power trains	Artificial teeth, bones, and joints	Memory element
Submarine shaft seals	Catalysts and igniters	Resistance heating element
Improved armors	Heart valves	Varistor sensor
Propulsion systems	Heat exchanger	Integrated circuit substrate
Ground-support vehicles	Reformers	Multilayer capacitors
Military weapon systems	Recuperators	Advanced multilayer integrated package
Military aircraft (airframe and engine)	Refractories	
Wear-resistant precision bearings	Nozzles	
Nuclear industry	**Oil industry**	**Electric power generation**
Nuclear fuel	Bearings	Bearings
Nuclear fuel cladding	Flow control valves	Ceramic gas turbines
Control materials	Pumps	High-temperature components
Moderating materials	Refinery heater	Fuel cells (solid oxide)
Reactor mining	Blast sleeves	Filters
Optical Engineering	**Thermal Engineering**	
Laser diode	Electrode materials	
Optical communication cable	Heat sink for electronic parts	
Heat-resistant translucent	High-temperature industrial furnace lining	
Porcelain		
Light-emitting diode		

6.6 Challenges in the Machining of Ceramics

There have been ample interests in possible and potential applications of ceramic materials during the last decades. These interests are followed by significant research and studies, which substantially promote the level of knowledge about ceramics. However, the challenges arise when it comes to manufacturing industrial components from ceramics where their attractive characteristics impose a huge barrier to their production. Several commonly used traditional machining operations, which work well in machining other materials, cannot be successfully implemented in machining ceramics. This is mainly due to the low fracture toughness and high hardness [16] of ceramics material which significantly elevate the cutting temperature at the cutting zone [17], increase the tool wear rate, and decrease the surface quality of the workpiece. This leads to high machining cost and confine the variety of shapes that could be made out of ceramics. For instance, milling and turning which are widely used machining operations for metals are not successful in the machining of ceramics especially when ceramics are fully sintered [17]. As a result, some time-consuming and costly machining operations, either traditional or non-traditional, are among the few options to machine ceramic components [18]. It must be noted here that ceramic components are normally used under the high mechanical and thermal load [18] where the surface quality and the level of residual stresses induced by machining play an important role in determining the performance during the service life. Therefore, those few applicable machining options must be vigilantly utilized to maintain the surface integrity of the workpieces.

Several studies have been performed to test the performance of various operations in cutting ceramic materials. Turning [17], grinding [19, 20], ultrasonic machining [18, 21, 22], abrasive water jet machining [23, 24], electrical discharge machining [25], laser machining [17, 18, 26], honing, lapping, and polishing [27] are among the machining operations that their capabilities in machining of ceramics have received utmost attention.

6.7 Mechanism of Chip Formation

In every machining operation, the quality of workpiece surface is affected to some extent by the chip removal process. Therefore, in order to understand the effects of brittleness in the machining of ceramic materials, the mechanics of chip formation must be thoroughly investigated. In machining of brittle materials, such as ceramics, chip formation can be attributed to the following two factors [28, 29]:

- Chip formation due to plastic deformation on the characteristic slip plane
- Chip formation due to cleavage fracture on the characteristic cleavage plane.

According to the above-mentioned factors, machining of brittle materials (crystalline materials) is more complicated than the homogeneous ones. When machining homogeneous materials, the plane at which maximum tensile or shear stress occurs matches the slip or cleavage plane. In contrast, during the machining of brittle (crystalline) materials, the chip can be formed by either plastic deformation of workpiece in front of the tool or by means of cleavage phenomenon [28, 30].

The plastic deformation occurs when the shear stress goes beyond a critical value on the slip direction before cleavage occurs. This critical value depends on the workpiece material properties. Alternatively, a cleavage takes place when the tensile stress perpendicular to the cleavage plane surpasses its critical value before the slip [28, 31, 32].

The mechanism of chip formation when machining ceramics is very important as it directly determines the surface characteristics of workpiece. A very smooth

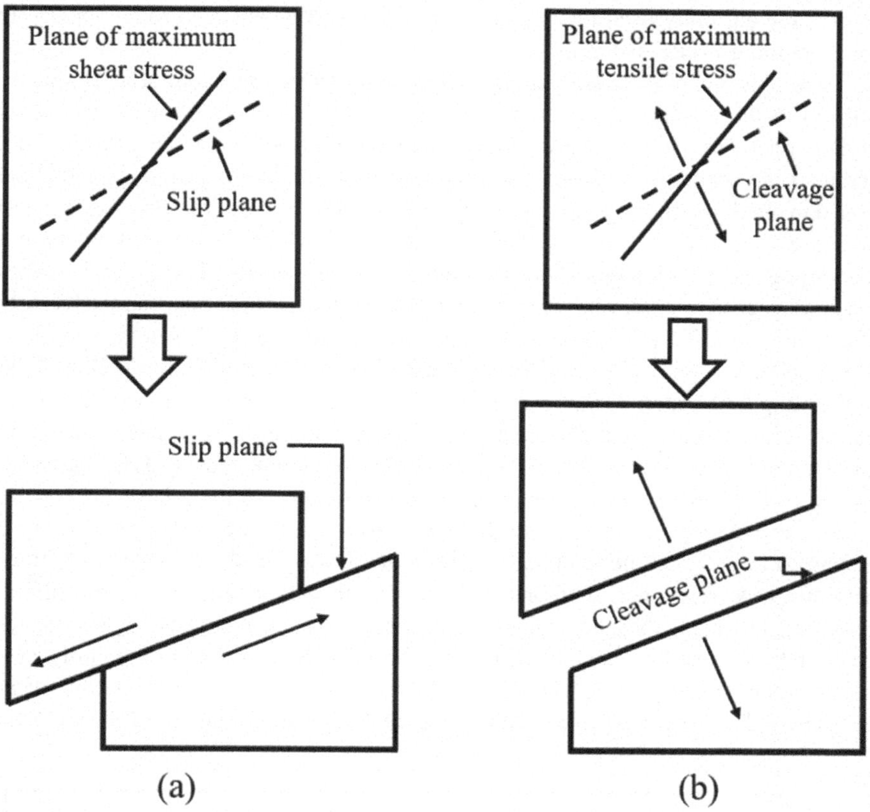

Fig. 6.1 Model of chip formation for brittle materials **a** plastic flow and **b** cleavage fracture (with permission to reuse) [28]

machined surface can be achieved if the material is removed by plastic deformation; however, the surface can be very coarse and dull with residual cracks if the cleavage fracture is the dominant mechanism of chip formation [28, 33]. The mechanism of chip formation is predominantly affected by cutting conditions. Machining brittle materials like ceramics can be carried out by controlling cutting conditions in a way so that the chip is removed from the workpiece surface by plastic flow rather than the cleavage fracture. This process is called ductile regime machining [31, 34]. Figure 6.1 demonstrates the model of chip formation for brittle materials.

6.8 Turning of Ceramic Materials

Although metals and metallic alloys, which are ductile or relatively ductile in comparison to ceramics, are widely used in different aspect of industry, brittle materials such as ceramics have also been extensively used for special purposes.

Hence, the currently used machining operations must be adjusted to be applicable for machining brittle materials.

Turning is one of the widely used machining operations that cannot be efficiently utilized in the machining of ceramic materials. High hardness of ceramics causes high temperature, which consequently increases the tool wear rate. In addition, ceramics are normally very brittle and the workpiece surface is extremely scratched and damaged by turning operation. One possible solution for this issue is to decrease ceramic hardness by thermal softening, which is referred to as hot machining [17]. Although hot machining softens the workpiece and makes machining easier, it elevates the temperature in the cutting zone and adversely affects the tool life by accelerating temperature-related wear [17]. Another option is to use an extra hard cutting tool that can sustain such high temperatures. One of the few cutting tool materials that can sustain such temperatures is diamond. Owing to its unique mechanical and chemical properties, a diamond tool is able to maintain an extremely sharp cutting edge even under elevated temperatures [31]. Figure 6.2 compares the hardness and wear resistance of different tool materials.

As can be seen in Fig. 6.2, the highest hardness and the best wear resistance belong to polycrystalline diamond (PCD) tools. Owing to their outstanding hardness and wear resistance, diamond tools are the primary candidates for machining ceramics; however, they are vulnerable to phase transformation at high temperatures and they can be transformed to graphite [17]. Next to PCD tools, polycrystalline cubic boron nitride (PCBN) tools have the highest hardness and wear resistance. They are also more stable at high temperatures up to 1400 °C. As a

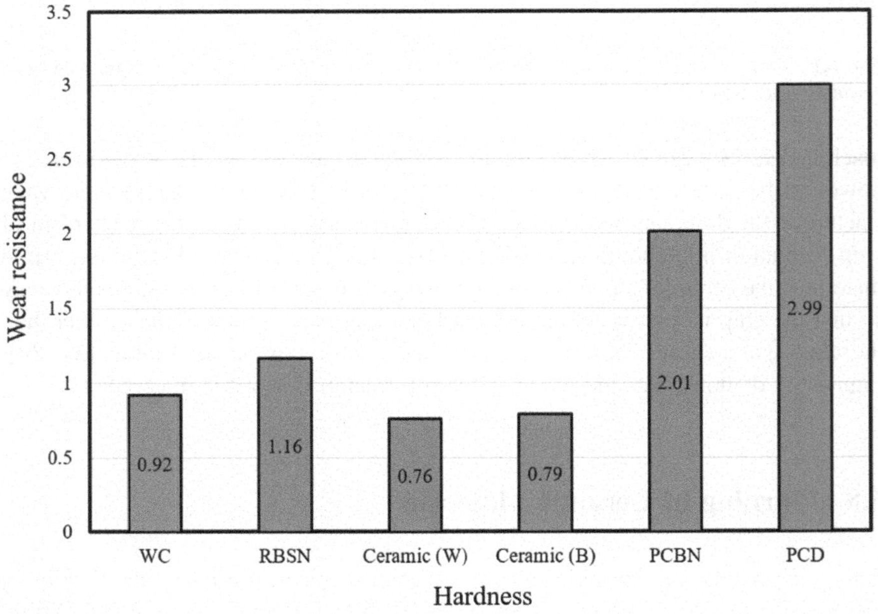

Fig. 6.2 Hardness and wear resistance of different tool materials (with permission to reuse) [17]

result, they can be considered as potential candidates for machining ceramic materials if the cutting temperature is controlled within a certain range [17].

Turning of brittle materials using diamond tool, generally known as diamond turning, is an essential process when spherical and non-spherical surface must be made with high quality without requiring further surface polishing processes [31]. Laser mirrors, polygon mirrors, and magnetic disks are good examples of such parts [28]. In order to achieve mirror surface finish, which is necessary in producing the above-mentioned parts in the optic industry, the material must be removed from the workpiece surface in ductile mode. It has been shown [28, 31] that the brittle–ductile transition border depends on several factors among them the following items are dominant:

- Material properties (crystal structure)
- Geometry of cutting tool
- Cutting conditions.

The border between brittle–ductile transitions is highly influenced by the crystallographic orientation of the workpiece material and the cutting direction [28]. Geometric features of cutting tool such as nose radius, rake angle, and clearance angle (relief angle) also affect the border between brittle mode and ductile mode machining. Small nose radius results in a small inference region between tool and workpiece, which consequently leads to a small size of critical stress field. When the stress field in front of the cutting edge is small, the ductile mode is more likely to occur. Rake angle and clearance angle also affect the stress distribution in workpiece material. To avoid brittle fracture, a negative rake angle is preferred due to imposing hydrostatic stress [28, 31]. Feed rate and depth of cut are also influential factors in achieving ductile mode while machining brittle materials. Small feed rate and depth of cut are preferred when machining brittle materials to minimize the interference region between tool and workpiece and consequently achieve ductile mode chip removal [28, 31].

6.9 Grinding of Ceramic Materials

Grinning is the most expensive machining operation if the unit volume of material removal is considered as the performance measure [35]. Despite several research works to investigate the alternative methods to substitute grinding as a costly and time-consuming operation, it is still one of the most widely used operations in the machining of ceramic materials. It has been reported that in some cases, 80% of the machining cost of ceramic parts directly belongs to grinding [35].

Grinding is normally considered as a finishing operation to obtain the required surface quality. However, when it comes to grinding ceramic materials, it can introduce defects in the form of scratches and microcracks to the workpiece at a scale that is tremendously hard to identify. These types of damages are very risky

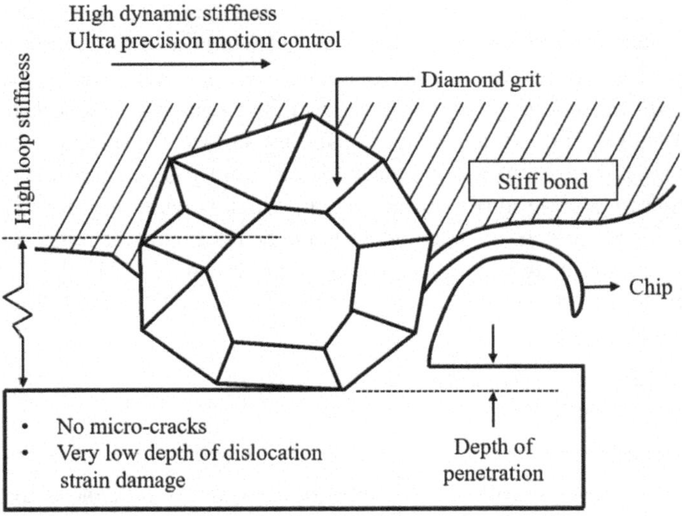

Fig. 6.3 Mechanism of ductile mode grinding of brittle materials (with permission to reuse) [31]

and impose many uncertainties about the strength of ceramic parts especially when they are used as structural components [36]. To eliminate these defects, the grinding of ceramics is normally followed by other finishing techniques such as lapping and polishing [18] which are also expensive and time-consuming.

The adverse effects of grinding can be effectively eliminated by controlling the cutting condition so that the grinding action takes place in ductile mode. A crack-free surface can be obtained if grinding wheel grits remove the material from ceramic surface by plastic flow instead of fracture. This process is called ductile regime grinding, and it is normally performed using a single-point or multi-point diamond tool [31]. Ductile regime during grinding of ceramic materials can be achieved when the amount of workpiece materials removed by each grit is small enough to let the brittle material plastically flow rather than fracture. Figure 6.3 illustrates the mechanism of ductile mode grinding of brittle materials.

Another problem during the grinding of ceramic materials is the high ratio of normal to tangential forces, which tends to elastically deform the structure of grinding machine as well as the grinding wheel. As a result, the grinding machine must be rigid enough to perform the task successfully [19, 35, 37, 38]

6.10 Ultrasonic Machining of Ceramic Materials

After grinding, ultrasonic machining (USM) is considered as the second most frequently used machining operation for ceramic materials. It is a non-traditional machining process during which a tool that has the inverse profile to the desired hole or cavity vibrates at high frequency (usually 20 kHz) and feeds toward the workpiece applying a constant force [18, 29]. The material removal action in ultrasonic machining is performed by a slurry consists of water and small abrasive particles, which is directly supplied to the cutting area between the workpiece and the tool tip [29]. The small abrasive particles, which are suspended in the slurry, hit the workpiece body by the down-stroke vibration of the tool and thus remove the material. Repeating this action consequently generates the desired shape. Therefore, some similarities can be distinguished between USM and grinding from microstructural viewpoint. Figure 6.4 shows the schematic illustration of USM.

The employed abrasive particles, which are suspended in the slurry, are undoubtedly an important factor in USM as they are directly responsible for micro cutting action. The frequently used slurry for ultrasonic machining of ceramic materials comprises of boron carbide as the abrasive particle suspended in water at a concentration of 20–50% [18]. The material removal rate and quality of generated surface are directly affected by the abrasive particles' size. Table 6.2 shows the attainable surface finish using different size of abrasive particles.

Ultrasonic machining can also be performed in the form of rotary ultrasonic machining (RUM) for machining difficult-to-cut materials. It can be effectively used for machining of ceramic matrix composites (CMC). This method is among the USM methods that can be used toward machining of ceramic materials [39, 40]. Despite regular USM in which the abrasive particles are suspended within the

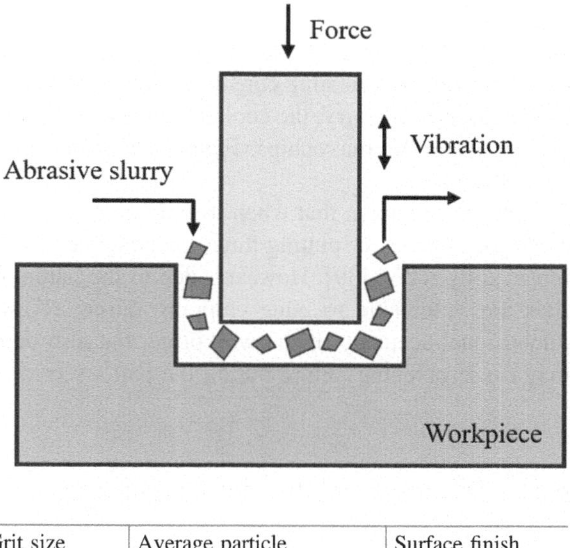

Fig. 6.4 Schematic illustration of ultrasonic machining (with permission to reuse) [29]

Table 6.2 Grit size and corresponding surface finish in ultrasonic machining (with permission to reuse) [18]

Grit size	Average particle size		Surface finish
	(mm)	(in)	(μin)
240	0.051	0.0020	25
320	0.038	0.0015	20
400	0.030	0.0012	16
600	0.020	0.0008	12

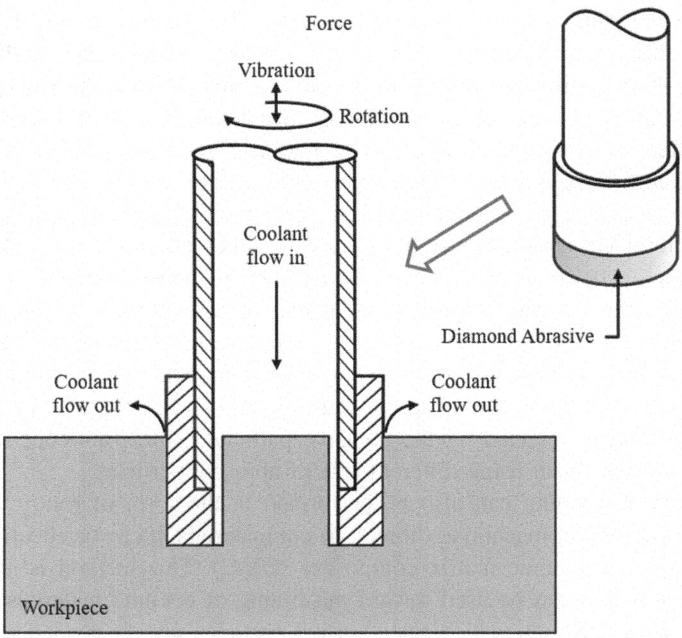

Fig. 6.5 Illustration of rotating ultrasonic machining (with permission to reuse) [39]

slurry, in RUM, the tool, which is ultrasonically vibrated and axially fed down to the workpiece, is a rotating core drill with metal-bonded diamond abrasive [39, 40]. In the absence of slurry, the coolant must be used to keep the cutting zone cool and also evacuate the microchips. Figure 6.5 shows the mechanism of rotating ultrasonic machining.

It has been shown that when machining ceramic matrix composites (CMC) is a matter of interest, the cutting force can be reduced to 40–60% if diamond drilling is replaced by RUM [39]. However, due to the nature of the process, ceramic materials are vulnerable to edge chipping during RUM which not only affects the dimensional accuracy of the workpiece, but also damages the surface quality and may cause potential failure during the part's service life.

6.11 Abrasive Water Jet Machining of Ceramic Materials

It has been stated earlier that the majority of traditional machining methods cannot be used in the machining of ceramics. The few that can be used, like grinding, are extremely time-consuming and costly as they involve the use of diamond tools, which are enormously expensive. Non-traditional machining techniques such as abrasive water jet machining (AWJ), electrical discharge machining (EDM), or laser cutting can be effectively used in the machining of ceramic materials. However, EDM and laser

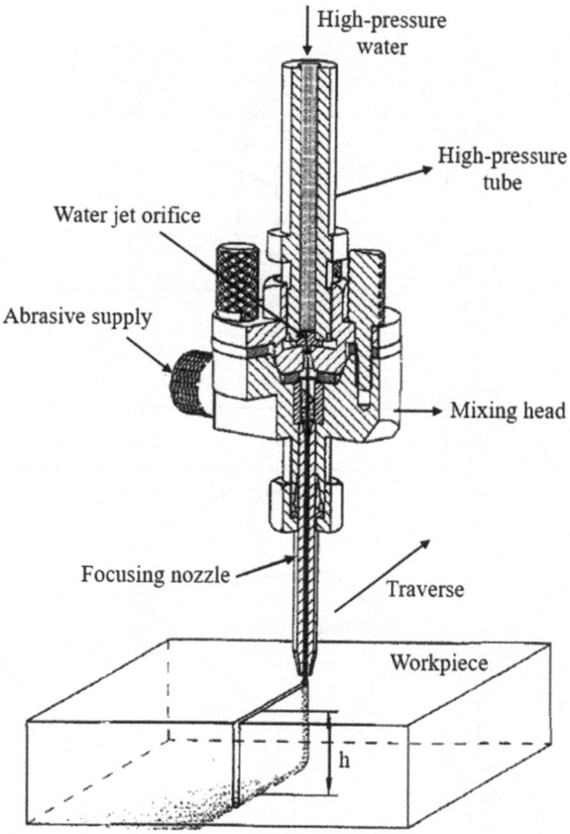

Fig. 6.6 Principle of the abrasive water jet machining (with permission to reuse) [43]

cutting are not able to cut relatively thick ceramics, and they may also generate heat affected zone which is not favorable [41]. In this case, abrasive water jet is the primary and the most promising candidate. Abrasive water jet machining (AWJ) is one of the non-traditional machining techniques in which a high-pressure flow of water, called water jet, carries abrasive particles. The workpiece is eroded by means of those abrasive particles, and the final shape is gradually produced. Figure 6.6 shows the principle of the abrasive water jet machining. AWJ dictates its superiority by offering the following advantages [41, 42]:

- No thermal distortion
- High flexibility
- High machining versatility
- Small machining force.

The performance of abrasive water jet machining is substantially influenced by the following factors [18]:

- Pressure of the water jet
- Distance between the nozzle tip and the workpiece surface (stand-off distance)

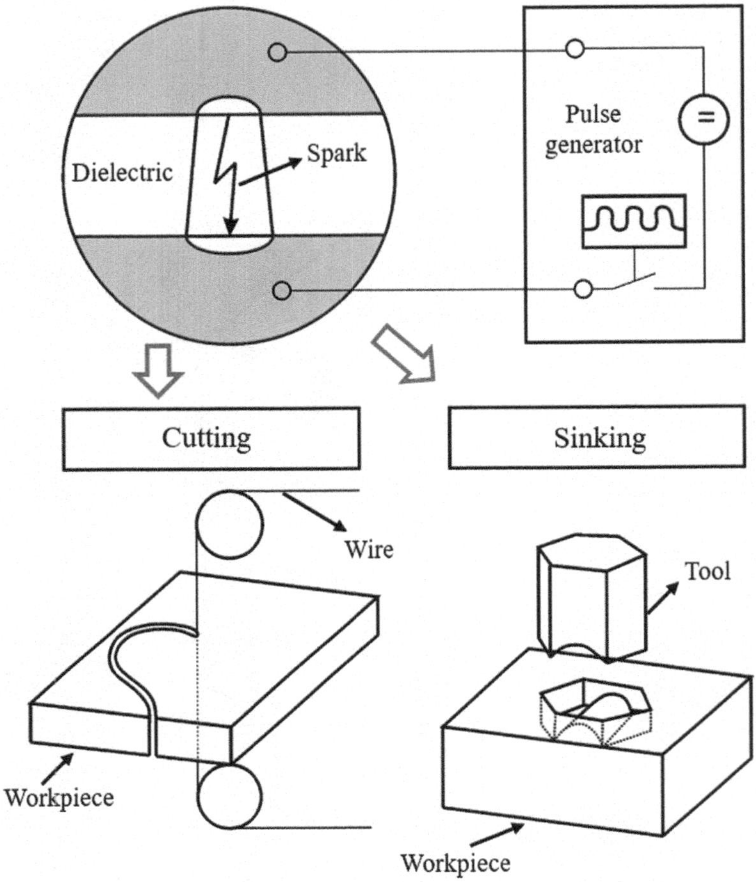

Fig. 6.7 Principle of the electrical discharge machining (cutting and sinking) (with permission to reuse) [45]

- Type of abrasive particles
- Size of abrasive particles
- Flow rate.

Among the above-mentioned important factors on the performance of the abrasive water jet machining, the characteristics of the abrasive particles are the most influential one. Abrasive particles such as garnet, alumina, or silicon carbide are normally used for machining low-strength ceramic materials among them garnet shows its significance by imposing the least wear to the nozzle [18]. Other abrasive materials such as alumina, SiC, or boron carbide are applied toward the machining of advance ceramics which are typically harder [18].

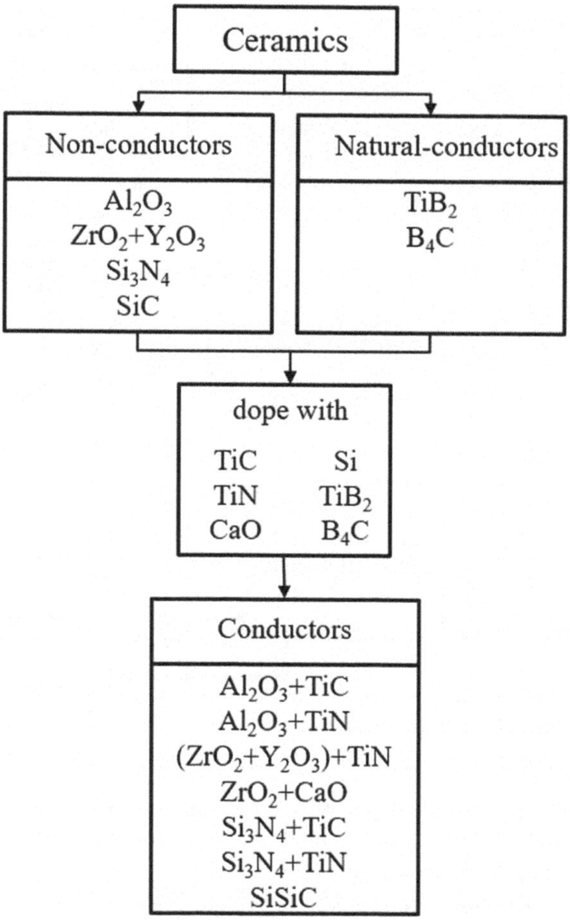

Fig. 6.8 Conductive and non-conductive ceramics (with permission to reuse) [45]

6.12 Electrical Discharge Machining of Ceramic Materials

Electric discharge machining (EDM) is one of the non-traditional machining processes in which the material is removed from the surface of the workpiece by means of electrical discharge or spark and the desired shape of the workpiece is eventually achieved [44]. In EDM, two electrodes, one of them is tool and the other one is workpiece, are separated from each other by a dielectric liquid and they are subjected to electric voltage. When two electrodes (tool and workpiece) are not too close to each other, the strength or electrical resistance of the dielectric does not let the electric current to flow between the tool and workpiece. However, when this distance is small enough, the intensity of the electric field between tool and workpiece (electrodes) becomes larger than the resistance of the dielectric, which allows the electric current to flow and initiate a spark. The electric spark rises the

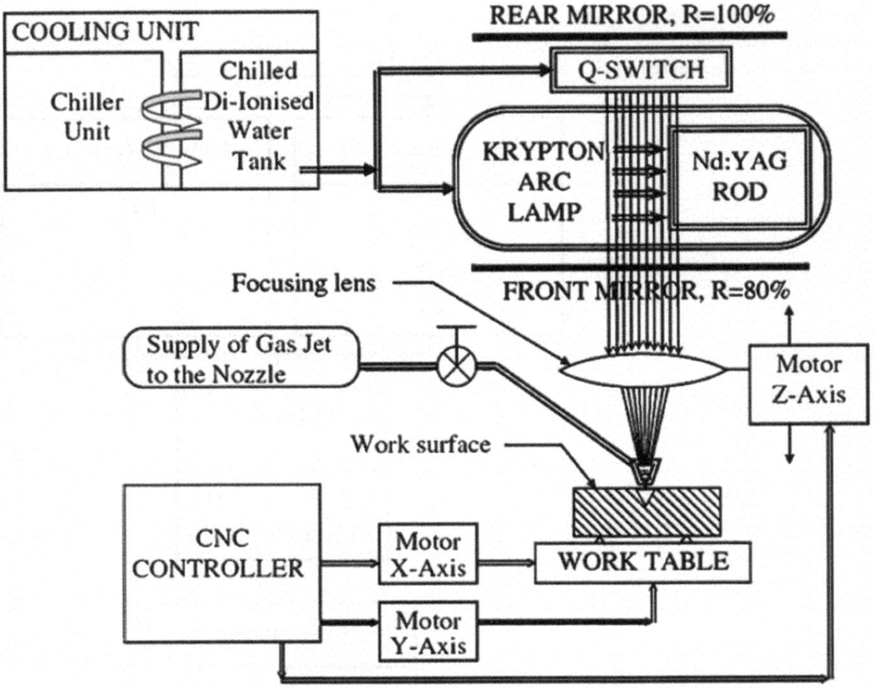

Fig. 6.9 Schematic illustration of laser machining (with permission to reuse) [51]

temperature locally and removes the material from both electrodes in micro scale. The EDM of ceramic material is generally performed in two ways: wire EDM or sinking, as shown in Fig. 6.7.

The only drawback in using the EDM in machining ceramic is that the ceramic being machined must be electrically conductive. A maximum electrical resistivity of 100 Ω cm is normally considered as the limit [45]. Some ceramics such as boron carbide (B4C) and titanium boride (TiB4) are naturally electrical conductive, and they have free lattice electrons to carry the electrical charges similar to metals [45]. If the ceramic that needs to be machined does not have sufficient electrical conductivity, two approaches can be undertaken to make it electrically conductive. These approaches are doping with conductive elements and incorporating impurity atoms [45, 46]. In the doping approach, the non-conductive ceramic is mixed with some conductive components during manufacturing. For instance, aluminum oxide (Al_2O_3) can be doped with 40% titanium carbide (TiC). Electrical conductivity can be also induced to ceramic by freeing additional electrons (incorporating impurity atoms) [45]. Figure 6.8 demonstrates the different categories of ceramics from the conductivity point of view.

During EDM, the material removal process is performed by means of two principal mechanisms [18, 47–49]. These two mechanisms are either melting (evaporation) or thermal spalling. The occurrence of each of these mechanisms depends on the properties of workpiece material and also the selected EDM parameters [49]. Although the melting is the most acceptable and common erosion mechanism during EDM, which results in smoother surface, the spalling can also be effectively used when more rapid and the efficient material removal rate is favorable. This is the case for rough machining of ceramics [49]. However, some substantial damage has been observed near the machined surface when the EDM parameters are set for maximum material removal rate. These damages may affect the service life of the part. Having several advantages over the other machining processes, EDM is an appropriate choice in machining ceramic materials. These advantages can be listed as follows:

- Ability to machine the workpiece with complicated shape
- Relatively high material removal rate
- Insensitivity to material hardness.

The only limiting factor that imposes a barrier to the application of EDM in machining ceramics is the electrical conductivity, which can be also modified in different ways as discussed earlier.

6.13 Laser Machining of Ceramic Materials

In laser machining, a high-density optical energy called laser (light amplification by stimulated emission of radiation), which is actually the source of energy in laser machining, is directed to the workpiece surface. The laser beam melts and then evaporates the workpiece and removes the material [50]. Figure 6.9 illustrates the schematic view of laser machining.

Laser machining has several significant advantages over the other methods that can be used in the machining ceramic materials. These advantages can be summarized as follows [50]:

- Some common issues in other machining processes such as the high magnitude of cutting forces and the consequent machine tool deflection, tool wear, and machine vibration are eliminated due to the nature of laser cutting where there is no physical contact between the machine and the workpiece.
- Laser machining depends more on the thermal characteristics of the workpiece material rather than its hardness; as a result, hard and brittle materials such as ceramics can be effectively machined using this technique.
- If laser cutting is combined and performed on CNC machines or multi-axis positioning systems, it can be flexibly utilized toward drilling, cutting, and grooving of ceramic materials without changing the machine.

The metal removal rate in laser machining of ceramic material is highly influenced by the material properties of the workpiece and also the laser processing parameters.

6.14 Application of Coolant in the Machining of Ceramics

A coolant is normally used during machining operations to lower the temperature at the cutting zone and minimize the negative effects of mechanical and thermal shocks to the machined surface. The coolant also keeps the cutting tool cool and decreases the rate of temperature-related wear.

The application of coolant in machining ceramic materials is included but not limited to some processes such as diamond turning, grinding, and rotary ultrasonic machining. Applying coolant in abrasive water jet machining, electrical discharge machining, or laser machining that can be used toward machining ceramics is not common. For instance, it has been reported [52] that the CBN tool wear can be significantly reduced while machining reaction bonded silicon nitride (RBSN) if liquid nitrogen is used as a coolant. As a coolant, the liquid nitrogen can decrease the CBN tool temperature from 1153 to 829 °C which is the safe temperature for this type of tools [52].

6.15 Concluding Remarks

Ceramic materials are generally used due to their significant characteristics such as high strength in compression that can be retained in high temperature, high hardness, great wear resistance and stability in corrosive environment. However, the superior characteristics that make ceramic an appropriate solution for different engineering problems also make them very challenging to cut. Here are the drawbacks:

- Ceramics are extremely brittle and have low fracture toughness. This makes them susceptible to failure due to the surface damages such as scratches and microcracks.
- High tool wear rate can be observed during machining ceramics primarily due to the high cutting temperature and extreme hardness.
- According to their special characteristics, many traditional machining operations that work well for other types of materials demonstrate poor performance in the machining of ceramic materials.
- Due to the incapability of regular machining operations in machining ceramics, some time-consuming and costly processes such as diamond turning and grinding are among the few options to machine the ceramic components.
- Non-traditional machining operations such as ultrasonic machining (USM), abrasive water machining (AEJ), electrical discharge machining (EDM), and laser cutting display a great capability in machining ceramic materials.

References

1. Oldfather WA. A note on the etymology of the word "ceramic". J Am Ceram Soc. 1920;3 (7):537–42.
2. Richerson DW. Modern ceramic engineering: properties, processing, and use in design, vol. 29. CRC press; 2006.
3. Cayne BS, Bolander DO. New Webster's dictionary and Thesaurus of the english language. Lexicon Publications; 1991.
4. Kingery WD. Introduction to Ceramics. 1960.
5. Carter CB, Norton MG. Ceramic materials: science and engineering. Springer; 2007.
6. Clyne T, Withers P. An introduction to metal matrix composites. Cambridge University Press; 1995.
7. Rawal SP. Metal-matrix composites for space applications. JOM J Miner Met Mater Soc. 2001;53(4):14–7.
8. Pask JA, Evans AG. Ceramic microstructures' 86: role of interfaces, vol. 21. Springer Science & Business Media; 2013.
9. Hein D, Hill G, Ramsay WRH. Raw or pre-fired: Kiln Construction at Sawankhalok, North Central Thailand, as a guide to ceramic history. Int J Hist Archaeol. 2004;8(4):247–66.
10. Ho J, Jow T, Boggs S. Historical introduction to capacitor technology. IEEE Electr Insul Mag. 2010;1(26):20–5.
11. Fujishima S. The history of ceramic filters. IEEE Trans Ultrason Ferroelectr Freq Control. 2000;47(1):1–7.
12. Agency UEP. Ceramic products manufacturing. 1996.
13. Ghandehariun A, Kishawy H, Umer U, Hussein H. Analysis of tool-particle interactions during cutting process of metal matrix composites. Int J Adv Manuf Technol. 2016;82(1–4):143–52.
14. Ducheyne P, Hastings GW. Metal and ceramic biomaterials. CRC Press Boca Raton; 1984.
15. Oonishi H, Aoki H, Sawai K. Bioceramics. Ishiyaku EuroAmerica; 1989.
16. Liang Y, Dutta SP. Application trend in advanced ceramic technologies. Technovation. 2001;21(1):61–5.
17. Wang Z, Rajurkar K, Murugappan M. Cryogenic PCBN turning of ceramic (Si_3N_4). Wear. 1996;195(1):1–6.
18. Tuersley I, Jawaid A, Pashby I. Review: various methods of machining advanced ceramic materials. J Mater Process Technol. 1994;42(4):377–90.
19. Nakagawa T, Suzuki K, Uematsu T. Highly efficient grinding of ceramics and hard metals on grinding center. CIRP Ann Manuf Technol. 1986;35(1):205–10.
20. Inasaki I. Grinding of hard and brittle materials. CIRP Ann Manuf Technol. 1987;36(2):463–71.
21. Kamoun H, Houlot M, Kremer D, Lecoco B. Modeling the material removal in stationary mode for ultrasonic contour machining. ASME Publ PED. 1993;64:759–759.
22. Thoe T, Aspinwall D, Wise M. Review on ultrasonic machining. Int J Mach Tools Manuf. 1998;38(4):239–55.
23. Savrun E, Taya M. Surface characterization of Sic Whisker/2124 aluminium and Al2o3 composites machined by abrasive water jet. J Mater Sci. 1988;23(4):1453–8.
24. Ramulu M, Hamatani G. Machinability of high temperature composites by abrasive waterjet. J Eng Mater Technol. 1990;12(4):381–6.
25. Martin C, Cales B, Vivier P, Mathieu P. Electrical discharge machinable ceramic composites. Mater Sci Eng, A. 1989;109:351–6.
26. Morita N, Ishida S, Fujimori Y, Ishikawa K. Pulsed laser processing of ceramics in water. Appl Phys Lett. 1988;52:1965.
27. Subramanian K. Precision finishing of ceramic components with diamond abrasives. Am Ceram Soc Bull. 1988;67(6):1026–9.

28. Nakasuji T, Kodera S, Hara S, Matsunaga H, Ikawa N, Shimada S. Diamond turning of brittle materials for optical components. CIRP Ann Manuf Technol. 1990;39(1):89–92.
29. Pei Z, Ferreira P, Haselkorn M. Plastic flow in rotary ultrasonic machining of ceramics. J Mater Process Technol. 1995;48(1):771–7.
30. Pramanik A, Basak A. Ductile mode turning of brittle materials and its practical aspects. In: Advanced materials research. Trans Tech Publications; 2013.
31. Ngoi B, Sreejith P. Ductile regime finish machining—a review. Int J Adv Manuf Technol. 2000;16(8):547–50.
32. Marinescu I, Rowe B, Ling Y, Wobker HG, Abrasive processes. In: Handbook of ceramic grinding & Polishing. Elsevier; 1999. p. 94–189.
33. Beltrão PA, Gee AE, Corbett J, Whatmore RW. Ductile mode machining of commercial PZT ceramics. CIRP Ann Manuf Technol. 1999;48(1):437–40.
34. Zhong Z. Ductile or partial ductile mode machining of brittle materials. Int J Adv Manuf Technol. 2003;21(8):579–85.
35. Malkin S, Hwang T. Grinding mechanisms for ceramics. CIRP Ann Manuf Technol. 1996;45 (2):569–80.
36. Zhang B, Zheng X, Tokura H, Yoshikawa M. Grinding induced damage in ceramics. J Mater Process Technol. 2003;132(1):353–64.
37. Mayer J, Fang G-P. Effect of grinding parameters on surface finish of ground ceramics. CIRP Ann Manuf Technol. 1995;44(1):279–82.
38. Inasaki I, Nakayama K. High-efficiency grinding of advanced ceramics. CIRP Ann Manuf Technol. 1986;35(1):211–4.
39. Li Z, Jiao Y, Deines T, Pei Z, Treadwell C. Rotary ultrasonic machining of ceramic matrix composites: feasibility study and designed experiments. Int J Mach Tools Manuf. 2005;45 (12):1402–11.
40. Li Z, Cai L-W, Pei ZJ, Treadwell C. Edge-chipping reduction in rotary ultrasonic machining of ceramics: finite element analysis and experimental verification. Int J Mach Tools Manuf. 2006;46(12):1469–77.
41. Siores E, Wong W, Chen L, Wager J. Enhancing abrasive waterjet cutting of ceramics by head oscillation techniques. CIRP Ann Manuf Technol. 1996;45(1):327–30.
42. Chen L, Siores E, Wong W. kerf characteristics in abrasive waterjet cutting of ceramic materials. Int Mach Tools Manuf. 1996;36(11):1201–6.
43. Momber A, Eusch I, Kovacevic R. Machining refractory ceramics with abrasive water jets. J Mater Sci. 1996;31(24):6485–93.
44. Jameson EC. Electrical discharge machining. SME; 2001.
45. König W, Dauw D, Levy G, Panten U. EDM-future steps towards the machining of ceramics. CIRP Ann Manuf Technol. 1988;37(2):623–31.
46. Lauwers B, Kruth J-P, Liu W, Eeraerts W, Schacht B, Bleys P. Investigation of material removal mechanisms in EDM of composite ceramic materials. J Mater Process Technol. 2004;149(1):347–52.
47. DiBitonto DD, Eubank PT, Patel MR, Barrufet MA. Theoretical models of the electrical discharge machining process. I. A simple cathode erosion model. J Appl Phys. 1989;66 (9):4095–103.
48. Patel MR, Barrufet MA, Eubank PT, DiBitonto DD. Theoretical models of the electrical discharge machining process. II. The anode erosion model. J Appl Phys. 1989;66(9):4104–11.
49. Trueman C, Huddleston J. Material removal by spalling during EDM of ceramics. J Eur Ceram Soc. 2000;20(10):1629–35.
50. Samant AN, Dahotre NB. Laser machining of structural ceramics—a review. J Eur Ceram Soc. 2009;29(6):969–93.
51. Kuar A, Doloi B, Bhattacharyya B. Modelling and analysis of pulsed Nd: yag laser machining characteristics during micro-drilling of zirconia (Zro_2). Int J Mach Tools Manuf. 2006;46 (12):1301–10.
52. Wang Z, Rajurkar K. Wear of CBN tool in turning of silicon nitride with cryogenic cooling. Int J Mach Tools Manuf. 1997;37(3):319–26.

Chapter 7
Environmentally Conscious Machining

Abstract The previous chapters covered different types of difficult-to-cut materials. Despite superior material properties and physical characteristics, the poor machinability is a common feature among all of these materials. Consequently, their exceptional characteristics and their candidacy as the paramount option for utilization in various applications are hindered by several challenges; most of them are directly associated with poor machinability. The main challenge is short tool life because of accelerated tool wear, which in turn lowers the productivity and increases the production cost. Other issues such as lack of dimensional accuracy and poor surface integrity are also commonly observed during machining difficult-to-cut materials. Among several factors that can be listed as the causes of poor machinability of some of these materials, the root cause is believed to be low thermal conductivity. This feature leads to the concentration of heat in the cutting zone and eventually rapid tool deterioration. In machining some difficult-to-cut materials, the heat induces other issues such as thermal errors and low dimensional accuracy. As a result, effective dissipation of heat from the cutting zone leads to improved machinability and better tool life. Heat dissipation in machining is commonly achieved by the utilization of cutting fluids. However, concern has recently grown due to the possible environmental and health hazards caused by cutting fluids. The growing concern has forced the governments and other associated agencies to impose strict policies and regulations to govern the application, recycle, and disposal of cutting fluids. Complying with governments regulations to reduce or eliminate the cutting fluids and thus minimize their associated hazards has dramatically increased the production cost. This is of particular importance when machining difficult-to-cut materials since excessive coolant is traditionally used due to the inherent characteristics of these materials. As a result, the industry aims at shifting from using flood cooling toward more economical yet environmentally friendly options. These options include minimum quantity lubrication techniques (MQL), environmentally friendly cutting fluids, nano-cutting fluids, self-cooling rotary tools, and eventually dry cutting. The current chapter presents a brief description of cutting fluids, different cooling strategies and their effectiveness, application of nano-cutting fluids in machining difficult-to-cut materials, and dry machining of these materials using self-propelled rotary cutting tools.

H. A. Kishawy and A. Hosseini, *Machining Difficult-to-Cut Materials*, Materials Forming, Machining and Tribology, https://doi.org/10.1007/978-3-319-95966-5_7

7.1 Introduction

During the process of chip formation, workpiece material undergoes severe plastic deformation because of tool penetration into the materials. A large amount of energy is usually consumed to cut the workpiece materials and form chips, and most of this energy is eventually converted to heat. In such a case, the temperature in the vicinity of tool tip increases as the operation continues until it reaches a steady state. Heat is mainly produced, particularly along the shear zone, due to the plastic deformation of workpiece material. Aside from the extensive plastic deformation, heat is also generated due to friction in two areas. These areas include chip–tool interface (along the rake face) and tool–newly machined surface interface (along the flank face). These surface rubbings occur under high normal forces, and heat is consequently generated. Therefore, in addition to the primary shear zone, the tool–chip interface and tool newly machined surface interface are the two other sources of heat generation in metal cutting. In this context, the tool–chip interface is usually referred to as secondary deformation zone and tool–newly machined surface interface is known as tertiary deformation zone.

The heat in the primary deformation zone is predominantly generated because of the elasto-plastic deformation of workpiece material. In the secondary deformation zone, the main mechanisms of heat generation are sticking–sliding friction (due to the motion of chip over the rake face of the cutting tool) along with plastic deformation.

Despite primary and secondary shear zones in which plastic deformation is a primary factor in generation of heat, the heat in the tertiary deformation zone is primarily produced by the elastic deformation of workpiece plus friction between back of the tool (flank face) and the newly machined surface. Figure 7.1 shows the three deformation zones in metal cutting operations.

The above-mentioned three zones of heat generation are common during machining almost the entire materials, but one should note that the amount of heat generated varies from one material to another. Process parameters such as cutting speed, feed, and depth of the cut play an important role in generation of heat during metal machining; among them, cutting speed is the dominant one while feed and depth of cut occupy the second and third place. It must be stated here that the partition of the generated heat is highly dependent on the process parameters. For example, at higher cutting speeds, larger amount of heat is carried away with the generated chips and thus much less heat is conducted through the workpiece. This promotes high-speed machining as a way to reduce the harmful effect of thermal expansion and its effect on the dimensional errors. In addition to machining parameters, the variation of the amount of generated heat also depends on the workpiece and tool material due to the difference in mechanical and thermal characteristics. For instance, materials with higher tensile strength generate greater amount of heat than those of lower tensile strength.

While heat can facilitate, to some extent, the cutting process by softening the workpiece material, its adverse effects on the other aspects of the process call for better control. The effect of temperature is significant, in some cases, and leads to

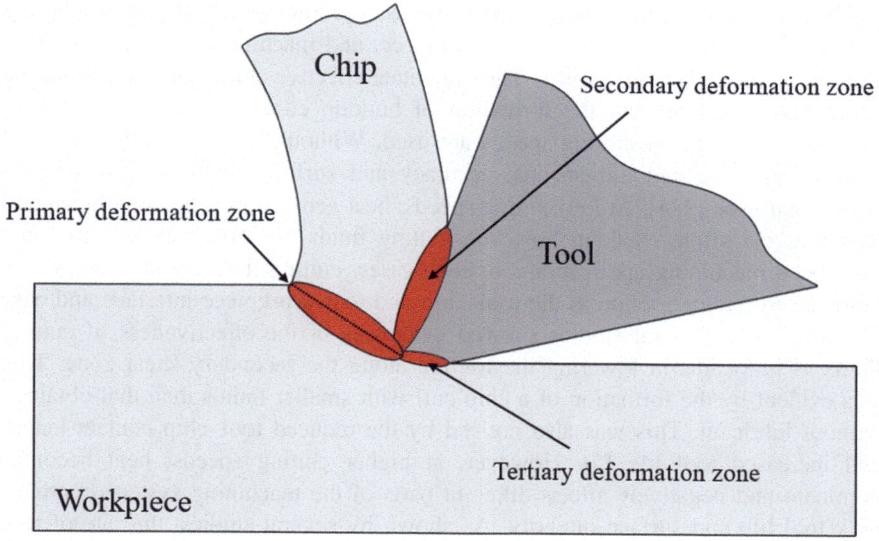

Fig. 7.1 Sources of heat generation in metal cutting

the deterioration of the tool integrity and also surface quality of workpiece. In addition, it leads to thermal expansion and thus dimensional inaccuracies. Elevated temperature also causes phase transformation and negatively affects the surface integrity of the workpiece. The most encountered surface integrity issue is the generation of tensile residual stresses, which negatively affects the fatigue performance of the machined parts. Another example of side effects introduced by heat is the formation of white layers on hard-turned surfaces. The white layer results from rapid cooling of the newly generated surface that leads to the formation of martensitic phase. As a result, extracting heat from the cutting zone and reducing friction are important considerations in any machining operation.

It is a common practice to use the term "coolant" in machining industry when referring to liquid agents, which are utilized to extract the heat and facilitate the process. This perception mainly originates from the fact that, at their early stages of evolution, these liquid agents were primarily implemented as a coolant. However, despite its widespread use, the word "coolant" may not be the right choice for this purpose and thus is more often replaced with the more meaningful term, "cutting fluid." Cutting fluid is a generic term that refers to several different products used in machining operations to satisfy two important requirements, namely cooling and lubrication [1]. Depending on the machining operation, one or both of these two requirements must be satisfied. As a rule of thumb, a cutting fluid must have great cooling properties when employed in machining operations such as turning and milling that are performed at relatively higher cutting speeds. On the contrary, lubrication becomes important in operations like broaching and tapping when cutting speeds are relatively low but cutting forces are high. In general, both functionalities are needed, with varying significance, in all cutting processes.

Cutting fluids are broadly utilized in industry, more specifically in machining industry, to protect the cutting tool, workpiece, and machine from excessive heat and thus extend their service life. They facilitate effective chip evacuation from the cutting zone and prevent the formation of buildup edge in most circumstances especially when lower cutting speeds are used. Without the proper application of cutting fluids, desired dimensional accuracy and surface quality cannot be efficiently achieved [2–4]. At low cutting speeds, heat generation is not significant and thus is not a prime concern; however, cutting fluids are still employed in some low-speed machining applications. In such cases, cutting fluid mainly serves as a lubricant to reduce friction at the tool–chip and tool–workpiece interface and ease machining [2]. Several studies showed evidences of the effectiveness of cutting fluids as lubricants in lowering the friction along the secondary shear zone. This was evident by the formation of a chip curl with smaller radius than that obtained without lubricant. This was also noticed by the reduced tool–chip contact length and increased tool life [5]. However, at higher cutting speeds, heat becomes dominant and negatively affects different parts of the machining systems, particularly tool life and surface integrity. As shown by several studies, the rate of tool wear increases because of elevated cutting temperatures generated at high cutting speeds. Under high temperature, the tool material loses its hardness and consequently loses its resistance to abrasive wear. In addition, the diffusion rate of elements from workpiece material into the tool material is higher at elevated temperatures. Thus, when higher cutting speeds are employed, the cutting fluids are more desired as a coolant rather than a pure lubricant to effectively dissipate the generated heat [2]. Needless to mention that cutting speed is categorized as low or high based on the combination of tool and workpiece materials and does not necessarily refer to a particular value.

In addition to cooling and lubrication as the two main required functions, cutting fluids must possess other characteristics. A successful cutting fluid must be nonflammable and pose no health hazard to the operator. It is also desired to be economical with no adverse effect on the machining system such as corrosion and/or discoloration of machine components, tool, and newly machined surface.

7.2 Traditional Cutting Fluids

Numerous types of cutting fluids are commercially available; however, they are not categorized under a universal classification. It is because of the fact that a single cutting fluid is often addressed by different names which lead to different types of classification. Some references categorize the cutting fluids into synthetic fluids, semi-synthetic fluids, emulsions, cutting oils, and gaseous fluids [1], while others classify them into neat or cutting oils, water-based fluids, gaseous fluids, air–oil mists, and cryogenic fluids [2]. Other classifications categorize the cutting fluids into two main categories of water-based or water-miscible fluids and neat cutting oils [3]. As mentioned above, despite being addressed by different names and

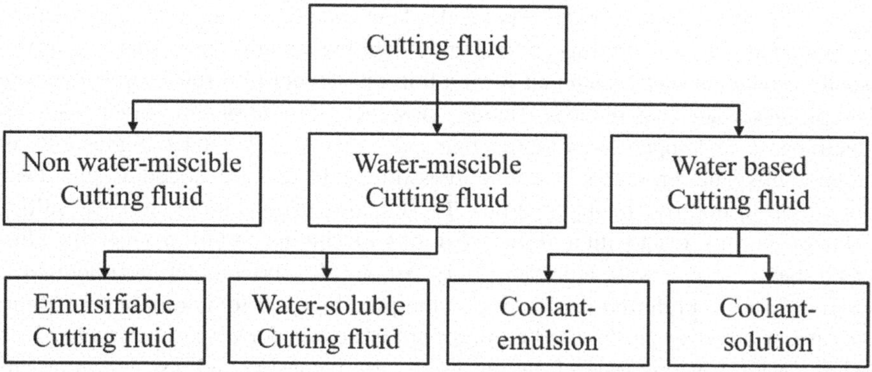

Fig. 7.2 Classification of cutting fluids according to DIN 51385 (with permission to reuse) [4]

categories, cutting fluids fundamentals are fairly simple and similar in all machining industries. In this chapter, widely used cutting fluids in the liquid state will be categorized according to DIN 51385 into [4]:

- Non-water-miscible cutting fluids
- Water-miscible cutting fluids
- Water-based cutting fluids.

The last category (water-based cutting fluids) is basically made by mixing the water-miscible concentrate to water [4]. Figure 7.2 shows three main categories of cutting fluids and their subcategories according to DIN 51385. It must be noted here that although cutting oils and water-miscible cutting fluids are by far the most widely used coolant/lubricant, liquids are not the only cutting fluids. Some materials in their solid or gas state can also serve the same purpose. Gases, pastes, waxes, and soaps are among the materials that are also used as coolant/lubricants during machining, some of which will be later discussed in this chapter.

7.2.1 *Non-Water-Miscible Cutting Fluids*

This category of cutting fluids is also known as neat cutting oils. The cutting fluids in this category are mostly mineral oils implemented without dilution with water [2] and are usually utilized as supplied without mixing with water. Other types of oils such as animal oils, vegetable oils (e.g., rapeseed, coconut, palm, and canola oil), and synthetic oils can also be categorized in this classification [2, 4].

Mineral oils are categorized as "straight" when they are used as the base oil without any additive. However, to increase the durability and improve lubrication, corrosion, and wear protection and their foaming properties, some additives such as vegetable oils (palm or rapeseed oil, castor oil derivatives), fatty animal oils (lard oil or tallow), and synthetic fatty substances (ester) are added with a concentration of 10–40% to the base mineral oils [2, 4]. This category of mineral oils is called compounded, and it is used more frequently than the straight ones. These additives usually produce a metal soap and form a thin lubricating film (half solid) over the workpiece surface that reduces friction. However, this lubricating layer loses its effectiveness in temperatures higher than 120–180 °C [4]. Under higher cutting temperatures and pressures, extreme pressure additives are added to the base mineral oil to improve their properties. These additives, including chlorine, sulfur, and phosphorous, react with tool surface during machining and form a metallic film, which then acts as a solid lubricant [2, 4]. Among the above-mentioned additives, chlorine is not a preferred choice due to the possibility of toxic dioxin formation. Dioxin is formed when the used cutting fluid is burned for disposal purpose. This makes disposal of this type of cutting fluid very expensive and also hazardous to both humans and environment. It must be noted that non-water-miscible cutting fluids or what are commonly known as cutting oils demonstrate great lubrication characteristics while their effectiveness as a coolant is not significant. Overall, based on their ingredients, non-water-miscible cutting fluids can be divided into the following subcategories each of which has its own applications, advantages, and disadvantages [6]:

- Mineral cutting oil
- Fatty oil and
- Blends of mineral oil and any of the following
  - Fatty oil
  - Sulfurized fatty oil
  - Sulfurized fatty oil and elemental sulfur
  - Sulfurized mineral oil
  - Sulfurized fatty oil and sulfurized mineral oil
  - Chlorinated paraffin
  - Chlorinated paraffin and sulfurized fatty oil
  - Chloro fatty oil
  - Chloro fatty oil and sulfurized fatty oil
  - Sulpho-chlorinated fatty oil.

7.2.2 Water-Miscible and Water-Based Cutting Fluids

As stated earlier, water-miscible cutting fluids and water-based ones are closely related, as water-based cutting fluids are simply a mixture of water-miscible concentrate and water. Therefore, in this section, these two categories of cutting fluids will be jointly discussed.

Water-based cutting fluids are mainly produced by dissolving or emulsifying oils in water. The proportion of oil to water is typically 1:10 or 1:60 [3]. Higher dilution can also be achieved by adding larger amount of water (up to 99%) [4]. A coolant is considered successful if it has high specific heat and also high thermal conductivity [6]. Water satisfies these requirements and fits the description of an effective coolant. It can dissipate the heat up to three times faster than mineral oils; also, water is relatively cheap, which gives it additional advantage. Water-based cutting fluids demonstrate great capability for cooling and chip evacuation, which make them a potential candidate for high-speed machining. However, the presence of water reduces their lubricity and promotes corrosion of workpiece and machine components, especially in the case of ferrous metals. Water also accelerates wear on the sliding and rotating surfaces of machine tools by washing away the lubricating oil. It also stimulates the microorganisms such as bacteria, yeasts, and fungi to grow [3, 4, 6], which leads to noxious odors and health hazards. To reduce such adverse effects, additives such as emulsifiers and inhibitors are usually added to the water-based cutting fluids in addition to their oil content. These additives increase lubricity, reduce corrosion, and avoid the growth of bacteria and fungi [3]. Water-miscible or water-based cutting fluids can be found in any of the following three forms [6, 7]:

- Emulsifiable oils (soluble oils)
- Chemical (synthetic) fluids
- Semi-chemical (semi-synthetic) fluids.

Emulsifiable oils or what are also referred to as soluble oils, emulsions, or emulsifiable cutting fluids [6] are emulsions (suspension) of oil, emulsifying agents, and sometimes other materials in water. Emulsifiers help forming a stable oil-in-water emulsion by breaking down the oil into tiny droplets and evenly dispersing them in water. Water-based cutting fluids in the form of emulsifiable oils offer excellent cooling properties owing to their water content. They also offer good lubrication and rust prevention due to their oil content. Combining cooling and lubricating properties together, emulsions are used for machining at high cutting speed (high heat) and low cutting pressure. They can be practically implemented in all light- to heavy-duty cutting operations, except for the heavy-duty machining of difficult-to-cut materials [6]. In comparison with straight or compounded cutting oils, emulsifiable oils are more effective coolants, are cleaner, and are economically feasible.

Some water-miscible or water-based cutting fluids are produced by dissolving inorganic and/or other materials in water. This category of cutting fluids that contains no mineral oil is called chemical or synthetic cutting fluids [1]. A majority of chemical (synthetic) cutting fluids are considered as coolant while few others demonstrate lubrication properties. This type of cutting fluids, especially those that contain no fat, is resistant against rancidity which is desired; however, their high detergency may irritate the skin after long exposure [1]. In addition, use of chemical cutting fluids containing nitrites is not recommended due to the possibility of health hazard.

The last category of water-based cutting fluids is the semi-chemical or semi-synthetic ones. This type of cutting fluids is mainly obtained by combining chemical fluids and emulsifiable ones; hence, they demonstrate some of the best characteristics of both categories. The oil content and thus the lubricity of these cutting fluids are lower than those of emulsifiable ones. However, since a small amount of mineral oil is added, they offer better lubricity in comparison with chemical (synthetic) cutting fluids [1, 6]. Moreover, adding different amount of extreme pressure additives positively alters the lubricating performance of semi-chemical cutting fluids. This category of cutting fluids can be utilized for moderate- to heavy-duty machining applications. Table 7.1 shows the different types of water-miscible (water-based) cutting fluids and their general characteristics.

Table 7.1 Water-miscible (water-based) cutting fluids (with permission to reuse) [6]

Class	Type and general characteristics
Emulsifiable oils	(1) General-purpose soluble oils Used at dilutions between 1: 10 and 1:40 to give a milky emulsion Used for general-purpose machining
	(2) Clear-type soluble oils Used at dilutions between 1:50 and 1:100 Their high emulsifier content results in emulsions that vary from translucent to clear Used for grinding or light-duty machining
	(3) Fatty soluble oils Used at similar concentrations to (1) and of similar appearance Their fat content makes them particularly good for general machining operations on nonferrous metals
	(4) EP soluble oils Generally contain sulfurized or chlorinated EP additives Used at dilutions between 1:5 and 1:20 where a higher performance than that given by (1), (2), or (3) is required
Chemical (synthetic) fluids	(1) True solutions Essentially solutions of chemical rust inhibitors in water Used at dilutions between 1:50 and 1:100 for grinding operations on iron and steel
	(2) Surface-active chemical fluids Contain mainly water-soluble rust inhibitors and surface-active load-carrying additives Used at dilutions between 1:10 and 1:40 for cutting and at higher dilution for grinding Most are suitable for both ferrous and nonferrous metals
	(3) EP surface-active chemical fluids Similar in characteristics to (2) but containing EP additives to give higher machining performance when used with ferrous metals Used at dilutions between 1:5 and 1:30
Semi-chemical (semi-synthetic) fluids	Essentially a combination of a chemical fluid and a small amount of emulsifiable oil in water forming a translucent, stable emulsion of small droplet size EP additives are usually included permitting their use for moderate- and heavy-duty machining and grinding applications

7.2.3 *Gaseous, Air, and Air–Oil Mists (Aerosols) Cutting Fluids*

Cooling and lubrication can be effectively accomplished when the cutting fluid deeply penetrates into the sources of heat generation. Owing to their ability to penetrate and cover all the cutting area, gaseous cutting fluids appear to be very attractive. They can penetrate deeply into the heat sources, e.g., tool–chip interface, and act as a heat sink. Moreover, in machining of reactive materials such as zirconium, using the liquid cutting fluids, especially the water-miscible or water-based ones, may contaminate the workpiece material. In such applications, air is usually used to prevent the side effects of liquid cutting fluids, especially the water-based ones. However, despite their great ability to penetrate deeper, gases have lower cooling capacity than liquids. But research works have shown that their cooling capacity can significantly increase when a jet of compressed air is aimed toward the cutting zone [1]. Another way to improve the cooling capacity of gaseous cutting fluids is by using cooled air [6]. Sometimes, a small droplet of water-based oil is mixed with high-pressure air to increase its lubricity. This type of cutting fluids is called air–oil mists or aerosols. In addition to air, inert gases such as nitrogen, argon, freon, helium, and carbon dioxide can also be used as gaseous cutting fluids.

7.2.4 *Cryogenic Cutting Fluids*

As mentioned above, gases are not as attractive as liquids when utilized as a coolant; hence, they are usually applied as a high-pressure jet at low temperature. If this temperature is very low (below −150 °C), the cutting fluid is called cryogenic cutting fluid. Cryogenic coolants with boiling temperature below −150 °C such as liquid nitrogen, liquid argon, and liquid carbon dioxide are occasionally utilized in machining difficult-to-cut materials like titanium and hardened steels [2] when extreme localized heat is a major concern. The cryogenic coolants are also used when machining other types of difficult-to-cut materials, e.g., stainless steels and nickel alloys; however, in such applications lubrication is also required and can be provided by an additional oil line [2] or solid lubricant.

7.3 Advanced Nano-Cutting Fluids

Cooling efficiency can be significantly increased by increasing the surface through which the heat is dissipated. In the context of metal cutting, effective heat dissipation remarkably improves the tool life and therefore production rate. One of the recently proposed cutting fluids is nano-cutting fluids that can be effectively implemented to face the heat dissipation challenges during machining processes.

Owing to their good heat extraction capabilities, nano-cutting fluids offer superior cooling characteristics [8]. In order to understand the mechanism of action when nano-cutting fluids are implemented, the term nano-fluid must be clearly explained. A nano-fluid can be produced by dispersing nano-additives into the base cutting fluid. The nano-additives can be nonmetallic, mixing metallic, carbon, or ceramic [9], and they are usually in the form of nanoparticles or nano-fibers (<100 nm) [10]. Using nano-fluids is advantageous due to the following reasons [11]:

- Larger surface to exchange heat between particles and fluids
- High dispersion stability
- Lower power consumption in the intensification of pure liquid
- Customized heat transfer properties and surface wettability by altering the nano-additives concentrations.

The cooling of electronic components, engines, nuclear systems, and ultimately machining operations are few examples among many where nano-fluids are employed due to their improved thermal, rheological, and stability properties [12–14]. Durability and stability are the two main factors that must be precisely controlled during the preparation of nano-cutting fluids to achieve optimum thermal properties. Between these two factors, stability is highly dependent on sedimentation velocity. Lower sedimentation velocity is always desired to ensure nano-fluid's stability.

Nano-fluids can be prepared using two different techniques. In one technique, nano-fluid is prepared by initially manufacturing the nano-additives and then dispersing them into the base fluid. In the other technique, both processes are performed simultaneously. The former (two-step) technique is simpler and mainly used for the dispersion of oxide particles and carbon nanotubes; however, issues such as nano-additives agglomeration can be a challenge. The latter technique is recommended for volume concentration values higher than 20% [15, 16]. Utilizing the single-step technique yields a stable and durable nano-fluid and prevents nano-additives agglomeration and sedimentation. Nevertheless, single-step technique is more complicated than the two-step one and is not suitable for the applications with high values of volume concentration [17, 18]. Regardless of the preparation technique, dispersion of nano-additives into the base fluid is a very important step as it significantly affects the viscosity and thermal conductivity of the nano-cutting fluid. To ensure uniform desperation, nano-additives are usually dispersed using an ultrasonic machine, which is then followed by a stirring stage using a magnetic stirrer. The processing time for producing nano-cutting fluids depends on the weight fraction (wt%) of nano-additives [19, 20].

7.3.1 Characterization and Performance of Nano-Cutting Fluids

Nano-cutting fluids can be characterized based on four design parameters including type of nano-additives (e.g., metallic/nonmetallic particles, carbon tubes, and graphene), type of the base fluid (e.g., water-based oil, organic liquids, vegetable oil), other additives (e.g., surfactants, antiwear/corrosion additives, and fungicides), and finally the scale, which is the percentage of weight-to-volume concentration [21]. When producing nano-cutting fluids, these design parameters are selected based on the desired functional requirements. The functional requirements are application dependent, which include thermal, tribochemical, physical, and rheological properties. Recently, researchers have focused on further improving the desired characteristics of nano-cutting fluids by adding multi-walled carbon nanotube (MWCNT) into the base fluid. Studies have shown that thermal conductivity can be improved up to 150–200% in comparison with that of the based fluid with no additive [22]. As previously mentioned, in addition to the cooling capability, cutting fluid must possess enough lubricity to reduce the friction, especially along the tool–chip interface. It has been shown that adding graphite nanoparticles to the conventional lubricants effectively reduces the friction and thus cutting forces. This ultimately yields better dimensional accuracy and surface quality [23]. Thermal conductivity can be further improved by increasing the concentration of nanoparticles in the base cutting fluid while decreasing their size [24].

7.3.2 Challenges in the Application of Nano-Cutting Fluids

Despite their superior properties, nano-cutting fluids are still facing several challenges, which hinder their wide implementation. Among these challenges, poor suspension stability is one of the main ones [13]. Because of strong van der Waals interactions in nano-cutting fluids, nanoparticles are easily aggregated. Although several solutions such as using surfactant have been proposed to address this issue, the time frame between preparing and utilizing the nano-cutting fluid is still a critical factor to consider. This is due to the fact that nanoparticles agglomerate over the time [25]. Other challenges in producing nano-cutting fluids and nano-additives are sedimentation and clustering. All of these challenges increase the production cost and make nano-cutting fluids relatively expensive [26]. Pressure drop and pumping power are other challenges that need to be addressed especially when nano-cutting fluids with higher viscosity and density values are utilized [13]. The thermal behavior of nano-cutting fluids in case of turbulent flow is another issue yet to be addressed. Few studies claimed promising results when using nano-cutting fluids in turbulent flow [27, 28]; however, making a consistent conclusion requires more research and investigation.

7.4 Delivery Methods of Cutting Fluids

Although the type and characteristics of a cutting fluid play an important role in its successful performance, the effectiveness of cutting fluids is considerably affected by the application or delivery method to the cutting zone. Needless to mention that the term "coolant" has been traditionally used for cutting fluids; hence, the methods of applying cutting fluids are still widely referred to as cooling techniques. The simplest and probably the oldest method of cutting fluid application is manual application with a small brush or a pump oil can [1]. These methods are effective for small jobs at very low cutting speeds, but they are considered outdated and not suitable in today's machining technology. In current machining market, four basic methods are usually employed to deliver coolants, namely low-pressure flood application, high-pressure flood application, high-pressure through-tool application, and mist application [2]. To employ these methods, most of the machine tools are equipped with built-in tanks or reservoirs to hold the cutting fluid and also a pumping system to deliver it to the cutting area.

7.4.1 Low-Pressure Flood Cooling

The most common method of applying cutting fluids is by low-pressure flooding. In this method, the cutting fluid is picked up from the reservoir and is delivered in the form of a steady stream by pointing a single, round nozzle toward the cutting zone area. Pressure for low-pressure flood cooling typically ranges from 5 to 20 psi [1], while the typical range of flow rate is between 10 l/min for single-point tools and 225 l/min for tools with multiple teeth such as milling tools [29]. This method has been widely used and works properly for machining operations such as turning and milling. In turning operation, the nozzle is positioned above the workpiece and it applies the cutting fluid downward. This single nozzle configuration aiming downward is not as effective as implementing two nozzles, one above and one below the cutting tool. The former only applies the cutting fluid over the rake face and chip, while the latter applies the cutting fluid to the flank face as well. Figure 7.3 shows the flood cooling from multiple nozzles for (a) milling and (b) turning to ensure that cutting fluid effectively reaches the desired areas.

Flood cooling is a common approach to ensure that cutting fluids continuously reach the cutting area; however, it drastically loses its effectiveness when higher cutting speeds are employed or more importantly when machining difficult-to-cut materials.

Insufficient amount of coolant or improper application leads to intermittent cooling and heating cycles, which are very damaging, and must be avoided. In milling operations, cyclic heating and cooling may lead to thermal fatigue wear of the milling tools. It has been shown by different studies [3, 30, 31] that in most of the machining operations under commercial cutting conditions, the cutting fluid

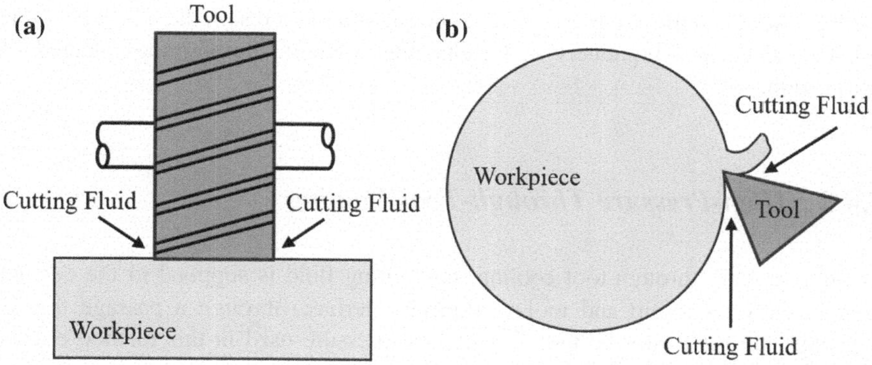

Fig. 7.3 Flood cooling using multiple nozzles in **a** milling and **b** turning operations

only penetrates up to the end of sliding zone where the chip starts to leave the rake face. It has also been demonstrated that the cutting fluid does not effectively reach the chip–tool interface especially the seizure zone close to cutting edge where most of the heat and friction is mainly encountered. In machining difficult-to-cut materials such as nickel and titanium alloys, cutting fluid only penetrates up to one-third of the overall chip–tool contact length [31].

7.4.2 High-Pressure Flood Cooling

One method to increase the efficiency of cutting fluid application is high-pressure flood cooling. It has been found that more effective cooling and lubrication and ultimately better machining performance can be achieved when cutting fluid is injected into the tool–chip interface at high pressure. High-pressure cooling technique applies the cutting fluid with high pressure through the nozzles pointed toward the cutting system. The nozzle pressure highly depends on the orifice diameter and varies between 5 and 100 bar (75–1500 psi) [2] although higher pressures have also been reported in the open literatures [32]. High-pressure cooling effectively removes the chips and lubricates the tool/chip/workpiece contact areas. However, generation of mist, which may be hazardous to the operator's health, is a major drawback that must be taken into account. High-pressure flood cooling can be performed in three different ways as follows [33]:

- Cutting fluid is injected into tool–chip interface by an external nozzle [34, 35].
- Cutting fluid is injected between the flank face and machined surface by an external nozzle [36].
- Cutting fluid is injected directly through the tool rake face into tool–chip interface [37–39].

The first two methods are similar to some extent as in both of them the cutting fluid is applied by means of an external nozzle to the desired area, but the last method uses the built-in passages in the cutting tool. As a result, the third method is usually categorized separately and called high-pressure through-tool method of application.

7.4.3 High-Pressure Through-Tool Cooling

In high-pressure through-tool cooling, the cutting fluid is supplied to the desired area (usually tool–chip and tool–workpiece interface) through a passage that is designed and built into the tool. The typical pressure used in this method ranges from 35 to 100 bar (500–1500 psi). One of the main advantages of this technique is that cutting fluid is directly applied over the desired area where maximum temperature is experienced. As a result, cutting fluid is not applied to any unnecessary areas; hence, less cutting fluid is required.

7.4.4 Mist Cooling

In the mist cooling technique, the cutting fluid is delivered by a high-pressure, high-speed stream of air in the form of mist [29]. The mist can be produced using either aspirator or direct pressure method [2]. In the former method, a stream of air is directed through an open tube containing oil to generate mist. This stream creates a partial vacuum and consequently pulls oil droplets from the container into the air stream. In the latter method (using direct pressure), mist is produced by directing compressed air through the oil. Mist cooling is mainly implemented for water-based cutting fluids and is highly advantageous for the applications where flood cooling methods are not practical, for instance, when the cutting zone cannot be reached easily by the cutting fluid [2]. It also leaves the workpiece visible during machining and gives operator more flexibility and control over the workpiece [29]. Despite its advantages, mist cooling may have adverse health consequences such as skin irritation or respiratory problem [40] if airborne fluid particles are inhaled; hence, proper ventilation and extra care and attention when using this method are necessary. The situation worsens in high-speed machining due to the likelihood of cutting fluid mist generation.

7.5 Cutting Fluids and Their Consequent Health Hazards

Application of cutting fluids is highly beneficial in metal cutting processes to ensure that machining of parts can be accomplished in a fast and cost-effective manner. However, achieving these goals is usually compromised by the health hazards and environmental side effects associated with the utilization of cutting fluids along with emerging governmental regulations. It has been shown that long-term exposure to cutting fluids either in liquid or mist form may increase the risk of health hazard [41]. Human body may absorb cutting fluids or their ingredients through dermal (skin) contact or inhalation, while other exposure mechanisms such as ingestion are less likely but possible. Microbial spoilage requires aqueous or partly aqueous environment; hence, substantial microbial fouling is possible when the nature of the job requires hand immersion or close proximity within the spray range, especially for water-based cutting fluids. The hazard risk significantly increases if the skin is damaged due to other external causes such as mechanical abrasion, open cuts, and chemical irritations [42]. The health hazards that may be encountered in long-term exposure are categorized into the following groups [2]:

- Toxicity
- Dermatitis
- Respiratory disorders
- Bacterial or microbial infections
- Cancer.

7.5.1 Toxicity

Short- or long-term exposure to either cutting fluids or their additives (e.g., biocides or fungicides) may induce toxic reactions in terms of acute or chronic toxicity [2, 43]. Acute toxicity mainly occurs due to unintentional ingestion through splashes, absorption through cuts, inhalation of mists, or even handling food with dirty hands, while chronic toxicity is known to be the result of long-term exposure to mists.

7.5.2 Dermatitis

Dermatitis is one of the most common adverse effects of exposure to cutting fluids. Dermatitis mainly refers to skin inflammation, but the term has also been widely used to denote a work-related skin rash with non-infective origin [41]. Skin can show two distinct reactions, namely direct reaction (due to contact) and indirect reactions (allergic) [2], when exposed to cutting fluids. Direct reaction can be initiated after prolonged exposure or minimal contact. For example, water is not usually damaging to the skin but it causes waterlogging after long-term immersion. Contrariwise, some other materials such as alkalis and solvents are harmful to the skin and initiate direct reaction after minimal contact immediately or in a very short period of time. Indirect reaction is mainly triggered by allergy, when the skin becomes sensitized and it reacts to a certain constituent of cutting fluid. The signs of reaction mainly occur in the vicinity of the point of contact; however, scattered rashes and swelling may also be observed in other parts of the skin, especially around the eyes. In occupational skin disease, irritant rashes most likely occur on the hand that is the main contact point with the cutting fluids [41]. Dermatitis can be simply prevented other than being treated. Providing machine operators with safety gloves, goggles, and hand creams, proper filtering of cutting fluids, and improving the education level on proper hygiene are common approaches to prevent dermatitis.

7.5.3 Respiratory Disorders

Another health hazard that may be encountered due to prolonged exposure to cutting fluid mists is acute respiratory disorders. These difficulties range from coughs, increased airway secretions, asthma, bronchitis, and airway constriction [2, 44–48]. These respiratory difficulties may eventually lead to shortness of breath. The situation is sometimes referred to as occupational asthma especially when the above-mentioned symptoms persist or worsen over time [2]. These difficulties are mainly related to the use of water-based cutting fluids that have the capability of producing mist, especially when used at high pressure [47]. Other work-related issues such as difficult breathing (dyspnea), fatigue, and hypersensitive pneumonitis (HP) have also been reported in workers who worked in machining industry where water-based cutting fluids were employed and recirculated under high pressure thus forming mist or aerosol [49]. Even though water-based cutting fluids are more prone to form mist and thus adversely affect the operators' health, oil mists and fumes can also be hazardous to those individuals who already suffer from asthmatic or chronic chest conditions [41].

7.5.4 Microbial Disorders

Another health concern raised due to implementing cutting fluids is microbial disorders. This issue is mainly associated with water emulsions, especially those produced by dissolving soluble cutting fluid concentrates in water to form emulsions. These water emulsions are intensively prone to microbial contamination even in the very normal machining conditions [50]. Pathogenic organisms such as klebsiella, staphylococcus, streptococcus, shigella, salmonella, and even dermatophytes have been repeatedly reported to be present in the cutting fluids by many research works and studies [42, 50–56]. Constant exposure to the contaminated cutting fluids may cause local skin infections especially if the skin is cut or wounded. It may also worsen other acute infections [50].

7.5.5 *Cancer*

As mentioned earlier, cutting fluids, especially water-based ones, are prone to degradation and microbial contamination. Growth of bacteria can be prevented by adding chemical additives to the cutting fluids. Chemical additives such as biocides and antimicrobials prevent the cutting fluid from contamination and help them to maintain their functionality [57]. However, these chemical additives introduce several health hazards to both environment and machine operators. Several currently used biocides release carcinogenic material known as formaldehyde [57]. According to the International Agency for Research on Cancer, prolonged exposure to cutting fluids containing mineral oils may cause occupational skin cancer due to the carcinogenic nature of the mineral oils [57, 58]. Biocides that are not releasing formaldehyde are also reported to be hazardous to the skin as they are highly corrosive [57, 59].

In some occasions, cutting fluids are utilized to suppress the solid dust generated during machining despite the fact that under high pressure and temperature the cutting fluids can be easily vaporized and form mist. It has been reported that the amount of airborne particles generated during wet machining can be 12–80 times higher than the amount of those generated during dry machining [57, 60]. These airborne particles stay in the air for a longer period of time and dramatically increase the chance of inhalation. Inhaling these particles could result in a wide range of diseases from mild respiratory problems to several types of cancers [57].

7.6 Environmental Considerations in Machining

Before the industrial revolution, the human activities, either industrial or agricultural, used to be performed at very small scales with almost no negative impact on the environment. However, the rapid development of industries after the industrial revolution and more specifically in the late twentieth and early twenty-first centuries swiftly changed the trend. Humankind started to conquer the nature by extracting the natural resources and implementing them toward what it was believed to be a dream life full of comfort. It took quite a long time until it was realized that rapid development of industry and unleashed exploration of natural resources have significantly harmed the environment. Humankind acknowledged the fact that his/her dream life could never be achieved without careful attention to the environment. Hence, reducing the adverse effects of technology on the environment became a priority although in practice it is not yet a top priority and prime concern in many industrial settings.

Environmental considerations in machining not only include concerns about application and recycling of cutting fluids, but also consist of several other factors such as energy consumption and carbon emissions. Hence, research topics such as green machining, environmentally conscious machining, environmentally benign machining, and sustainable machining have made their way to the list of the most attractive research topics in both academia and industry. These terms are usually used interchangeably; however, they convey different meanings that must be clearly differentiated.

Among the above-mentioned terms, sustainable manufacturing is the most comprehensive one. It aims to develop economically efficient manufacturing technologies yet addressing environmental concerns by minimizing the emission of greenhouse gases, minimizing the generation of waste, and encouraging the use of renewable and biocompatible materials. Sustainable manufacturing combines environmental, economic, and social concerns together with the main objective to address them simultaneously [61, 62]. Green manufacturing is a manufacturing strategy that is mainly focused on reducing the consumption of natural resources and minimizing the negative impacts on the environment. In other words, green manufacturing aims to minimize waste and pollution [61]. Environmentally conscious manufacturing is mainly focused on environmental footprints of manufacturing industry by minimizing the discharge of pollutants and reducing the production of by-products during the manufacturing processes [61].

Cutting fluids are among the most undesirable yet very important components in machining industry. As stated earlier, cutting fluids primarily serve two main purposes, namely cooling and lubrication. They are widely used all around the world during machining operations to extract heat, reduce friction, and help the cutting tool to survive longer and perform more effectively. They also help achieving better surface quality and dimensional accuracy. However, global awareness about the side effects of using cutting fluids on the operators' health raised several concerns among the end users on how to reduce the risks by eliminating or minimizing the application of cutting fluids. On the other hand, environmental concerns regarding recycling and disposal of cutting fluids constantly compel the governments and other governing agencies to enforce stringent legislations pertaining the application of cutting fluids. These two limiting factors are the main driving forces in the search for cutting fluids, cooling strategies, and machining techniques with minimal threat to the environment without compromising the quality of the process.

Environmental and health concerns can be satisfactorily addressed by directing the machining strategies toward minimizing, or in the ultimate case eliminating the use of cutting fluids. These two tactics introduced two relatively new machining approaches, namely minimum quantity lubrication (MQL) and dry machining. In the following sections, MQL and dry machining and their potential applications in machining of difficult-to-cut materials will be presented.

7.6.1 *Machining with Minimum Quantity Lubrication (MQL)*

Despite traditional methods where the tool–workpiece interface is flooded by cutting fluid (coolant, lubricant, or both), minimum quantity lubrication (MQL) applies only a small amount of cutting fluid, mainly with the lubricating action, directly to tool–workpiece interface. MQL is known as near dry lubrication or microlubrication [29, 57]. Implementing MQL significantly reduces the usage of cutting fluid and thus minimizes the environmental impacts. It also reduces the complexity of machining system and machine tools by eliminating equipment such as big cutting fluid reservoirs, circulators, and filters that are necessary in flood cooling to treat the cutting fluid and keep it functional.

In minimum quantity lubrication, a mixture of very small amount of cutting fluid and air, in the form of a very fine mist, is sprayed directly toward the cutting zone [29, 57]. This mixture covers the tool–workpiece interface with a thin film of lubricant and significantly reduces the friction as one of the primary sources of heat generation during machining. The thin film of lubricant evaporates at the cutting zone; therefore, maintenance and disposal of cutting fluid are not a prime concern in minimum quantity lubrication. Needless to mention that MQL is primarily a lubricating method rather than a cooling technique; however, if properly implemented, it has been proven successful in reducing the heat generation through minimizing the friction. The flow rate of cutting fluid in MQL has been reported to be in the range of 10–100 ml/h [57, 63]. However, flow rates in the range of 20–100 ml/h [61] and 50 ml/h up to 2 l/h [29] have also been reported.

Depending on the requirements of the process, cutting fluids are traditionally selected based on their primary characteristics such as cooling capability (for coolants), lubricity (for lubricants), and a combination of them for those cutting fluids that have served both functions. Although lubricity is still regarded as a very important feature, cutting fluid in MQL must possess other key characteristics such as long-term storage and oxidation stability as well as biodegradability [64]. The cutting fluids used in MQL must be biodegradable or compatible with the environment because MQL is mainly introduced to address the environmental concern. As a result, vegetable oils are normally used for MQL applications. Cutting fluid must also remain chemically stable under long-term storage due to the fact that the consumption rate of cutting fluid in MQL is very low, sometimes one ten-thousandth of that in flood cooling [29, 65]. Thus, cutting fluids used in MQL may be stored for a long period of time. Due to significant reduction in the consumption of cutting fluid, MQL guarantees longer equipment life, dryer and cleaner chips, less energy consumption, and finally reduced cost of recycling and disposal of the cutting fluid.

As stated earlier, the MQL is a lubricating method; hence, it usually offers poor cooling performance. In order to promote the cooling performance of MQL technique without violating the environmental concerns, a minimum amount of cutting fluid with cooling capability has also been implemented. This technique usually is

Minimum Quantity Cooling Lubrication (MQCL)
Normal consumption per machine hour:
10 – 50 ml MQCL medium

Minimum Quantity Cooling (MQC)	Minimum Quantity Lubrication (MQL)
Emulsion (water + oil) Water $C_{p,water} = 4.18 \frac{kJ}{kgK}$ Air $C_{p,air} = 1.04 \frac{kJ}{kgK}$	Oil $C_{p,oil} = 1.92 \frac{kJ}{kgK}$

Medium	Cooling	Lubrication	Chip removal
Emulsion	excellent	good	excellent
Oil	good	excellent	good
Air pressure	little	no	little

Fig. 7.4 Minimum quantity cooling lubrication (MQCL) and its media (with permission to reuse) [4, 64]

referred to minimum quantity cooling (MQC). MQL and MQC can be utilized simultaneously to achieve the best possible machining experience by reducing friction and extracting heat. The combination of minimum quantity lubrication and minimum quantity cooling is known as minimum quantity cooling and lubrication (MQCL) [66]. The generally used medium in MQCL is straight oil; however, emulsions or water has also been used as the media. The fluid media can be transported to the cutting zone directly or by means of a transport medium such as air [64]. Figure 7.4 shows the different media in MQCL and their typical characteristics.

7.6.2 Dry Machining

The economic aspects of using cutting fluids in machining industry have radically changed in the recent years due to the strict regulations imposed on the application and recycling the cutting fluids. These regulations were put in place mainly because of the environmental side effects and health hazards pertaining to the application of cutting fluids. Enforcing the environmental guidelines and complying with the regulations increased the manufacturing cost and put a burden on the manufacturers and customers. Although an accurate value for the cost associated with the application of cutting fluid is not available, estimations show that this cost encompasses at least 16% of the end-user manufacturing cost; see Fig. 7.5 [67].

Although 16% may not be considered as a significant contributor in the first glance, in the competitive manufacturing market it has a huge potential for reducing the overall cost. The potential opportunity to reduce cost through minimization or elimination of cutting fluid persuaded/forced the manufacturer to search for machining strategies and techniques with minimum or zero cutting fluid consumption. The evolution of relatively new techniques such as minimum quantity lubrication (MQL) and minimum quantity cooling lubrication (MQCL) was the result of such efforts. MQL and MQCL methods introduced several advantages over the traditional methods of cutting fluid application. A proper application of these methods significantly reduces the cutting fluid consumption and directs the trend toward more environmentally and user-friendly machining operations. Even though MQL and MQCL use minimum possible amount of cutting fluid, they still pose hazards to the operators' health when cutting fluid is applied directly or sprayed in the form of aerosol to the cutting zone. The cutting fluid gets vaporized especially when sprayed; thus, it can be inhaled by the operator [61]. Therefore, the ultimate goal each machinist may be dreaming for is to eliminate the application of cutting fluids entirely to not only address the environmental and health concerns but also to reduce cost and save money.

Dry machining, also known as ecological machining, refers to a cutting process in which no cutting fluid is utilized. It is the best approach to satisfy a wide range of concerns such as recycling and disposal of cutting fluids, environmental pollution (water and air), and ultimately operators' health safety. The main advantages of dry machining can be listed as follows [61]:

- Possible water pollution and soil contamination during the disposal process of cutting fluid are eliminated.

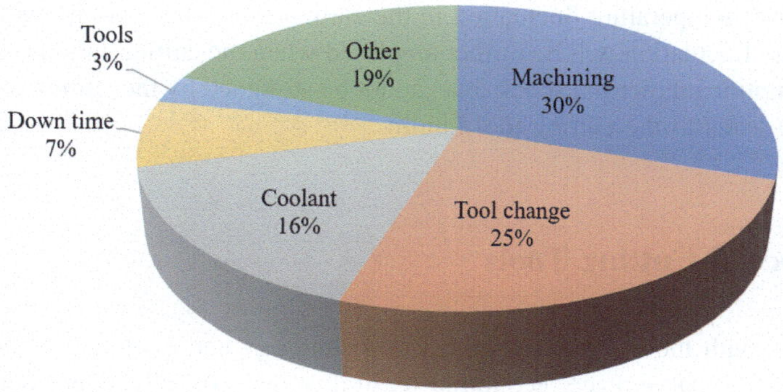

Fig. 7.5 Typical end-user manufacturing costs [67]

- The swarf (solid waste including chips, metallic dust, and small metal pieces) ends up dry; thus, no further chemical treatment is required during recycling to remove cutting fluid residue.
- Skin injuries and possible allergies due to contact with cutting fluids are avoided.
- Cutting fluid related costs including the cost of purchasing cutting fluid itself, proper storage and maintenance, and labor will be entirely saved.
- The cutting forces are sometimes reduced, especially in high-speed machining, due to the softening effect of high temperature on the workpiece material at the cutting zone.
- Absence of cutting fluid reduces the thermal shocks, mainly in interrupted machining operations such as milling; thus, it improves the tool life to some extent.

Despite the great advantages that make dry machining a very promising approach, several issues associated with dry machining hinder its widespread application. These issues are directly related to the excessive temperature generated because of plastic deformation in the primary shear zone and friction at chip–tool and tool–workpiece interface. This is a major concern in machining difficult-to-cut materials; for instance, it is impossible to dry machine titanium. In the absence of cutting fluid, the lubricating function of cutting fluid is not available; hence, more heat is usually generated during machining. Moreover, this heat cannot be effectively dissipated as the cooling function of the cutting fluid is also missing [68]. Consequently, the heat intensifies and damages both cutting tool and workpiece. Excessive tool wear and inferior surface finish are commonly observed problems in dry machining that significantly increase the machining cost and reduce productivity [67].

Based on the above-mentioned limitations, it can be concluded that among many machining operations, some of them are more adaptable to dry machining. Machining operations in which the tool is not enclosed by the workpiece are the primary candidates for dry machining. Milling operation, especially those of open surfaces, can be effectively performed in the absence of cutting fluid. During milling operation, the tool engages and disengages the workpiece in each rotation, which causes temperature fluctuation in the cutting tool. The undesirable effects of temperature fluctuation will be further amplified when the cutting fluid is used. This imposes significant thermal shock that cannot be sustained by the cutting tool, and it eventually leads to the cutting tool failure.

7.7 Special Cutting Tools

Nowadays, with the swift development of technology and evolution of new competitors in the marketplace, economic feasibility and cost-effectiveness in manufacturing are more important than ever before. Constrained by several limitations such as available budget, operating costs, quality requirements, and more importantly environment compatibility, achieving economic feasibility is not an easy target to reach. To achieve these ambitious business goals simultaneously, new technologies are being driven to their limits. New problems and challenges arise in par with the evolution of new technologies each of which requires a novel solution. As mentioned earlier in this book, one of the main challenges machining industry constantly encounters is machining the materials to be used in harsh service environments. These materials usually demonstrate unique characteristics that make them safe options for severe service conditions. However, these favorable characteristics also make these materials notoriously difficult to cut. They exhibit one or a combination of features that are all considered unfavorable from machining perspective. These features include but are not limited to abrasiveness, high hardness, hot hardness, high strength, low thermal conductivity, and low modulus of elasticity. These features drastically reduce tool life and thus increase the tooling costs and downtime for tool replacement. The rapid tool wear, which is caused directly or indirectly by temperature, blunts the cutting tool until holding part specifications within allowable tolerances become impossible. Slower cutting speeds and feed, especially for finishing cuts, accompanied by the application of cutting fluid are usually the remedies for this issue. However, lowering cutting speed and feed increase the cycle time and decrease productivity. In a mass production setting, longer cycle time leaves a dramatic impact on overall costs.

One of the primary responses to the demand for more durable cutting tools was developing tool materials with superior wear resistance capable of machining at high cutting speeds and sustaining higher temperatures. Tool materials such as coated carbide, ceramic, CBN, PCBN, and PCD are among the materials that were developed to address this need. However, the demand for more reliable cutting tools with longer service life did not only encourage the development of new tool materials, but also inspired geometric modifications and novel tool designs. One of the novel yet simple ideas to reduce the temperature-related issues and protect the cutting tool from rapid deterioration, especially when machining difficult-to-cut materials, was to replace the active portion of the tool or, in other word, by constantly providing a fresh cutting edge to the cutting zone [69]. Looking impracticable at the first glance, it is the fundamental idea behind the introduction of a certain generation of cutting tools named as rotary tools.

7.7.1 Self-propelled Rotary Tools

In its simplest form, a self-propelled rotary cutting tool is a round insert capable of freely rotating about its axis. The rotation of insert replaces the in-cut portion of the cutting edge with a fresh portion. Figure 7.6 shows the typical configuration of a rotary tool in turning and milling operation.

One of the pioneering works about the application of rotary tool in turning operation was done by Shaw et al. They presented a disk-shaped tool for turning

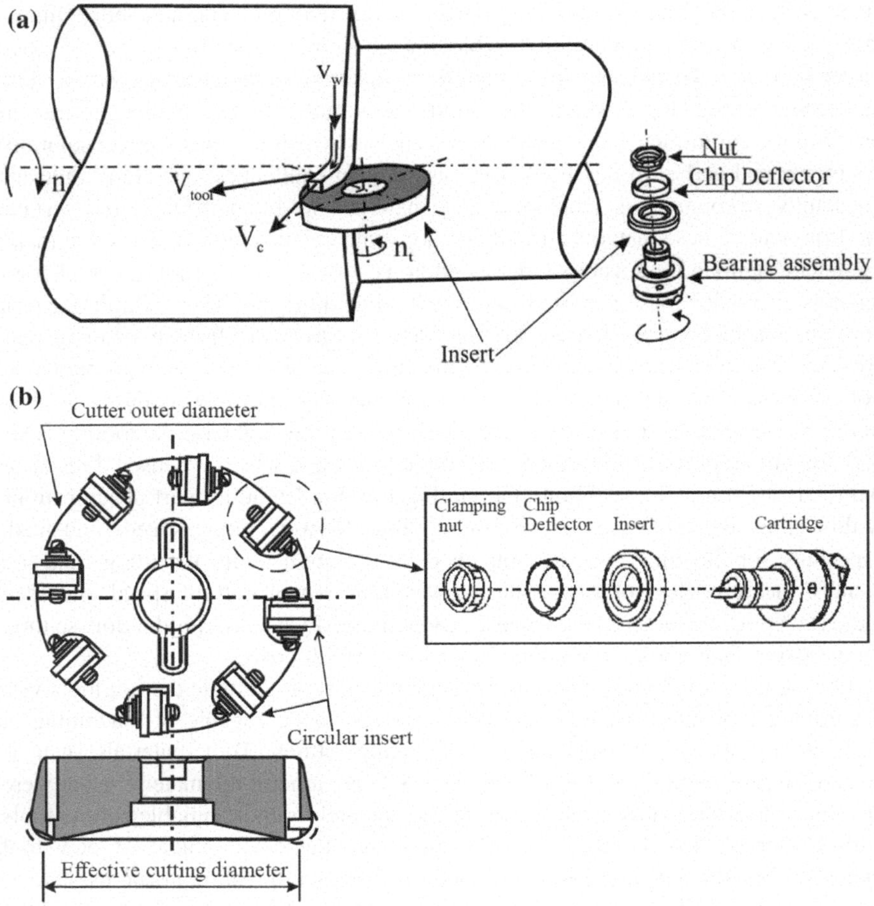

Fig. 7.6 Typical configuration of a rotary tool in **a** turning and **b** milling operation (with permission to reuse) [70, 71]

operation capable of rotating around its center [72]. This work paved the way for further studies about rotary tools and their applications in machining industry [71, 73–77]. Performance of self-propelled rotary tools can be comprehended through comparison with their stationary counterparts. Analyzing the machining operation using self-propelled rotary tools reveals significant differences when compared to the traditional tools used in machining industry. Among these differences, the type of relative motions between the tool and the workpiece stands out. In conventional machining operations, the cutting action is performed by a combination of cutting and feed motions. These two motions are generated by either rotating the workpiece and moving the tool (turning) or rotating and moving the tool simultaneously (milling). One may be tempted to consider milling tools as rotary tools; however, although milling tools are rotating during the process, they are not considered as rotary tools in this context as there is no relative motion between tool itself and its holder. In addition, the milling cutter is driven by the spindle while the insert in self-propelled tools is driven by the friction along the chip tool interface. A close analysis shows that the rotation of tool in milling operations is exactly similar to the rotation of the workpiece in turning. In both scenarios, the rotation provides cutting motion; therefore, the contact area along the cutting edge remains in cut for a period of time until the tool leaves the workpiece. When the tool is continuously in cut, the contact area is exposed to elevated temperature. This elevated temperature is mostly localized, particularly when machining difficult-to-cut materials. Continuous exposure to high temperature deteriorates the tool and contributes to the development of crater wear on the rake face due to constant rubbing of the chip against the tool. It also accelerates flank wear because of ploughing contact between the flank face of the tool and the newly machined surface [69, 70].

In conventional tools, there is no relative motion between the cutting inserts and their holder while in self-propelled rotary tools cutting inserts freely rotate around their central axis. Based on the source of motion for the circular insert, rotary tools are categorized into two main types, namely driven and self-propelled [74, 75]. Driven rotary tools are forced to rotate by means of an external source such as motor or actuator, whereas self-propelled rotary tools (SPRT) are driven when they engage with the workpiece and the tool motion occurs in the absence of any external actuator. Study of the insert rotation in self-propelled rotary tools revealed its dependency not only on the cutting speed but also on the type of material from which the insert is made; see Fig. 7.7.

The angular (rotational) velocity of the insert depends on the combination of insert/workpiece materials due to the relative coefficient of friction. Measuring the angular velocity of insert for different tool materials when cutting heat-treated AISI 4340 (54–56 HRC), hardened steel revealed that carbide tool rotates faster than coated carbide and ceramic. It was also observed that the dependency of angular velocity of insert to the combination of tool/workpiece material is more significant

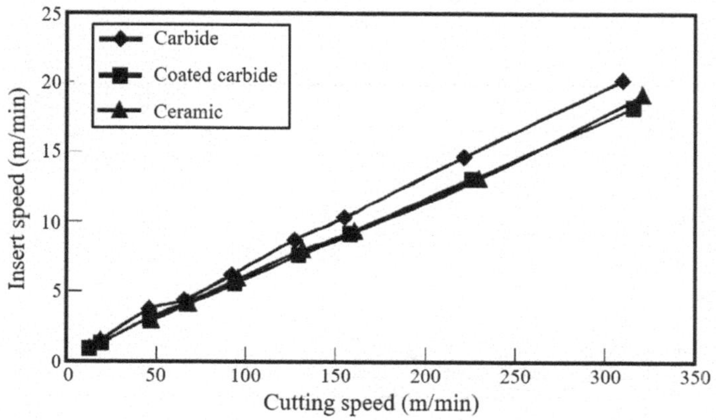

Fig. 7.7 Dependency of insert speed on tool material when machining heat-treated AISI 4340 (54–56 HRC) (with permission to reuse) [70]

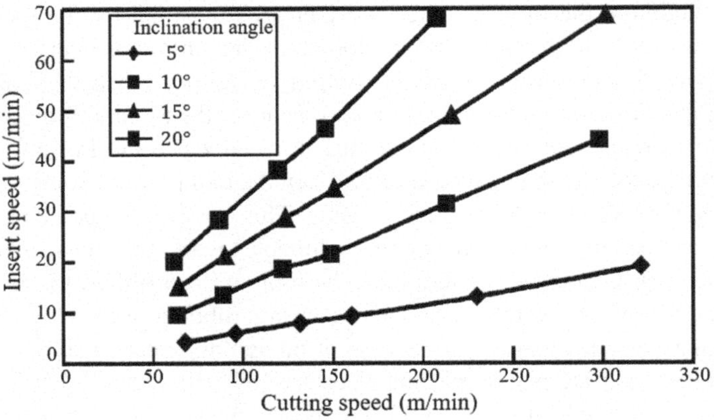

Fig. 7.8 Dependency of insert speed on tool inclination angle when machining heat-treated AISI 4340 (54–56 HRC) (with permission to reuse) [70]

at higher cutting velocities [70]. The tool inclination angle was also found to be another parameter that influences the insert rotation as shown in Fig. 7.8.

7.7.1.1 Self-Cooling Feature of Rotary Tools

Due to their inherent functional characteristic, during a full rotation of the insert, each portion of the cutting edge performs cutting for a short period of time and has a relaxation period until it reaches the cutting area again. The periodic engagement of each portion of the insert circumference allows for the self-cooling of that potion once it exits the cutting area. The self-cooling feature of the rotary tools was investigated in [70, 78] by modeling the effect of tool speed on the heat partitioning, R_2, along the chip–tool interface. This is illustrated by the rapid reduction in R_2 when the velocity ratio ($V_{tool}/V_{workpice}$) initially increases as shown in Fig. 7.9.

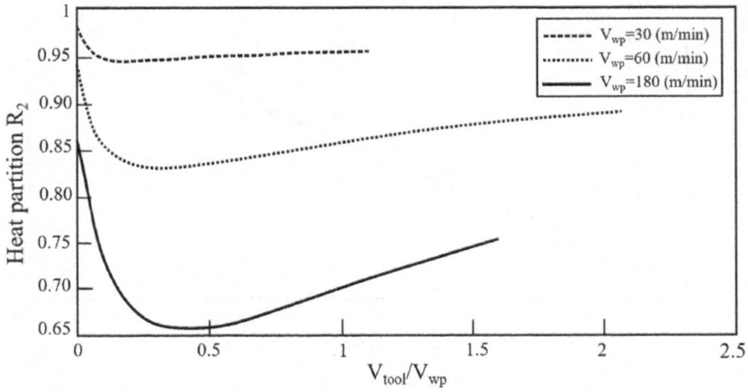

Fig. 7.9 Effect of insert rational speed on the amount of heat carried away by the insert (with permission to reuse) [70, 78]

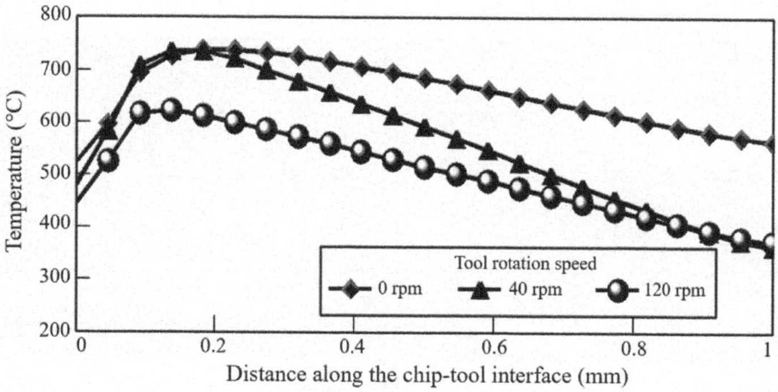

Fig. 7.10 Temperature distribution along the chip–tool interface when using self-propelled tool (with permission to reuse) [78]

It can be seen that there is an optimum insert rotational speed beyond which the portion of the insert that leaves the cutting area does not have enough time to cool down prior to returning to the cutting area again [70].

Study of temperature distribution on the tool rake face shows that increasing the tool rotational speed reduces the peak temperature [70]. As shown in Fig. 7.10, increasing the insert rotational speed does not only influence the peak value, but also shifts the position of the peak temperature toward the cutting edge. The self-cooling feature reduces the insert average temperature and thus enhances its resistance to diffusion wear.

The spinning of cutting tool not only prevents the prolonged exposure of a certain point of cutting edge to workpiece, but also carries the cutting fluid to the tool point at high cutting speeds [70]. Allowing continual lubrication to the chip–tool interface significantly reduces the harmful effect of friction at the secondary zone. In dry cutting, as a more environmentally friendly approach, rotation of cutting insert offers a self-cooling feature. The tool rotation dissipates the accumulated heat to the surrounding environment and allows the tool face to be cooled down during the relaxation period. This exactly resembles a tool with several cutting edges over which the tool wear is uniformly distributed. This unique configuration significantly reduces the tool wear and increases the material removal rate [79].

7.8 Machining Titanium and Superalloys Using Rotary Tools

As mentioned earlier in this book, extreme temperatures accompanied by excessive tool wear are common challenges when machining difficult-to-cut materials. Excessive heat becomes even more challenging during finish machining of titanium alloys and nickel-based superalloys as it softens the tool material and changes the geometry of cutting edge. Heat also promotes certain wear mechanisms such as

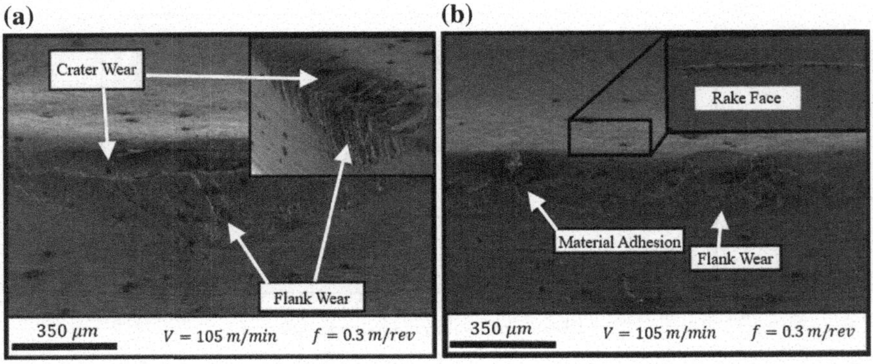

Fig. 7.11 Modes of tool wear when machining Ti-6Al-4V with **a** conventional carbide insert and **b** self-propelled rotary carbide insert (with permission to reuse) [69]

diffusion, which in turn accelerates crater wear and flank wear and negatively affects the integrity of the machined surface. Although the adverse effect of heat can be reduced by using ample amounts of cutting fluid, environmental considerations and strict health regulations make this approach unfavorable [69]. Another approach is using rotary tools with minimum amounts of cutting fluid or without cutting fluid. Figure 7.11 shows typical modes of tool wear observed when machining titanium (Ti-6Al-4V) with traditional and self-propelled rotary carbide tool.

The tool wear progression during machining titanium (Ti-6Al-4V) using conventional (fixed) tool versus self-propelled rotary tool (both carbide) is presented in Fig. 7.12.

As can be seen in Fig. 7.12, the rotary insert yields a significantly longer tool life based on the prescribed 0.3-mm flank wear criterion. This can be attributed to the use of entire circumference of the cutting insert in rotary tool. Machining nickel-based superalloys using rotary tools yields the same result [69]. Analyzing

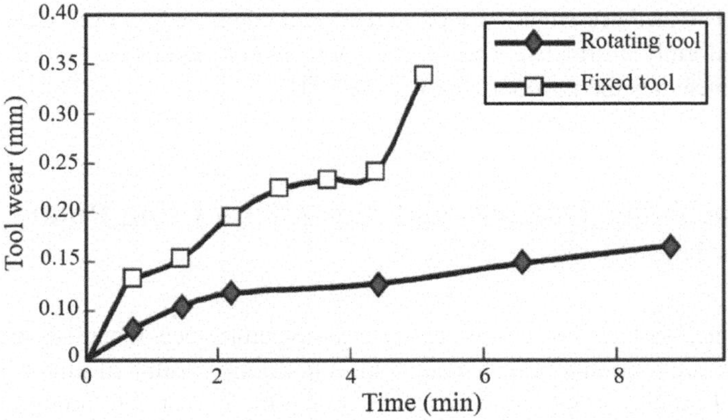

Fig. 7.12 Tool wear progression when machining Ti-6Al-4V with fixed and rotary carbide insert V = 105 m/min, f = 0.3 mm/rev, d = 0.2 mm (with permission to reuse) [69]

the modes of tool wear during machining titanium with standard carbide tool showed crater wear as the dominant mode of tool failure (see Fig. 7.11a), whereas evenly distributed flank wear is the dominant mode of tool failure when rotary insert of same tool material is implemented (see Fig. 7.11b). Machining nickel-based superalloys using fixed and rotary carbide inserts shows the same pattern of tool wear [69]. During the rotation of insert, the currently engaged portion of the cutting edge is continuously replaced by a new portion; hence, the adverse effect of heat is diminished and thermally induced failure becomes less critical and the dominant mode of tool wear changes from crater to flank wear [78].

7.9 Machining Hardened Steels Using Rotary Tools

Grinding is a common abrasive machining process to achieve high-quality surfaces; however, due to its inherently low metal removal rate, it is considered as a slow and time-consuming process. Hence, as previously shown, hard turning of hardened steels was developed as a better alternative. In addition, grinding is usually performed in the presence of plenty of cutting fluid, which makes it not an environmentally friendly process. Dry hard turning not only reduces machining time and the specific cutting energy, but also eliminates the health hazards and environmental concerns associated with the application of cutting fluids. Although dry hard turning is an interesting alternative for grinding from both economic and environmental perspectives, it introduces two interrelated challenges in terms of the tool wear and surface integrity of the machined workpiece [70]. A suitable cutting for hard turning must demonstrate high temperature and high wear resistance, and it must also be strong enough to endure high cutting forces that may be encountered during hard turning [70]. Moreover, due to the high hardness of the workpiece, ceramic and PCBN tools are primary candidates for hard-turning operations while carbide single-point cutting tools are not recommended [79]. Studies showed that during hard turning of AISI 4340 steels (35–60 HRC) using CBN tools, the hard carbide particles present in the workpiece highly abrade the tool binder material. Ceramic tool was also reported to dominantly suffer from adhesion and abrasion wear mechanisms [80].

Implementing rotary cutting tools during dry hard turning offers promising results. Due to their inherent self-cooling feature, rotary tools can sustain severe machining conditions compared to conventional non-rotating counterparts. The rotational feature of insert distributes the wear evenly over the entire length of the cutting edge and allows commercial carbide tools to be employed during hard turning. This makes the process more economically feasible without sacrificing the surface integrity and tool life as the carbide tools are much cheaper than ceramic and CBN tools [79]. Figure 7.13 shows modes of tool wear for a fixed and a rotary carbide tool after hard turning of AISI 4340 (54–56 HRC) under the same cutting condition.

It can be seen from Fig. 7.13 that the fixed carbide tool failed catastrophically, while self-propelled rotary tool of the same carbide experienced gradual flank wear. As can be seen, severe crater wear, which is the commonly observed mode of wear

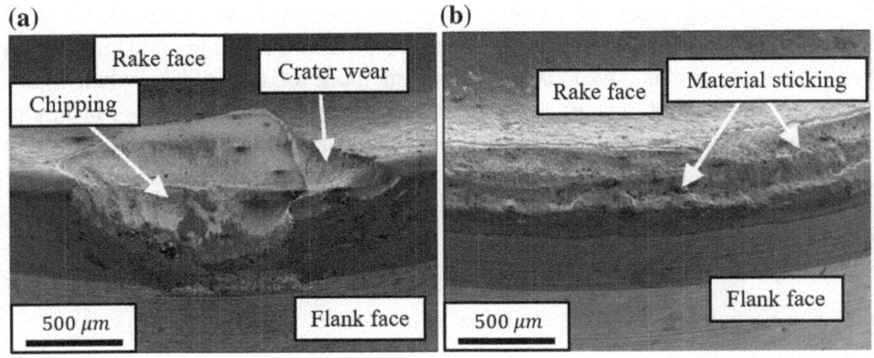

Fig. 7.13 Modes of tool wear when machining AISI 4340 hardened steel (54–56 HRC) with **a** fixed and **b** rotary carbide tool (with permission to reuse) [79]

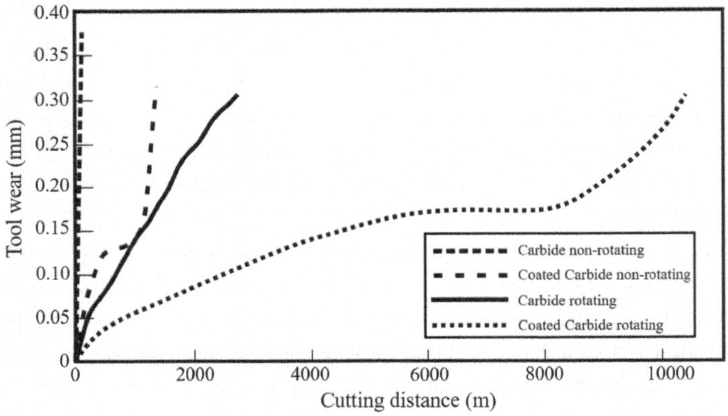

Fig. 7.14 Progress of tool wear for different tools (V = 120 m/min, f = 0.2 mm/rev, d = 0.1 mm) (with permission to reuse) [70]

during hard turning, does not exist in the rotary tool. This can be attributed to the self-cooling feature that effectively dissipates the heat as discussed earlier. The progress of tool wear during hard turning using self-propelled tool and traditional tool is illustrated in Fig. 7.14 using coated and non-coated carbide inserts. The improvement of tool performance is very evident. The tool life of coated carbide inserts increased by about ten times when self-propelled tool is used.

7.10 Concluding Remarks

Applications of cutting fluids and environmental concerns in machining industry, particularly in machining of difficult-to-cut materials, have been discussed in this chapter. Cutting fluids, in terms of coolant, have been proved very practical during machining to extract the heat from the cutting zone and reduce the adverse effects of heat on the cutting tool as well as on the workpiece. In terms of lubricant, cutting fluids effectively reduce the friction and thus facilitate the machining and also reduce the heat. However, application of cutting fluids in machining is always accompanied by several concerns such as cost, maintenance, recycling, and health effects among which the adverse effects on health is the prime concern. In order to address these concerns, the industry trend is toward minimizing and if possible eliminating the use of cutting fluids. However, minimizing or eliminating the use of cutting fluids make cutting tools prone to plastic deformation and premature failure due to excessive heat, especially during machining of difficult-to-cut materials. In this regard, new generation of cutting tools, known as rotary tools, has been introduced to the market. These tools rotate about their axis during machining; as a result, each portion of the cutting edge is in cut for a short period of time and leaves the workpiece thereafter. Leaving the workpiece periodically allows the tools to relax and effectively dissipate the heat to the surrounding atmosphere. This self-cooling feature extends the tool life and reduces the tool cost. Cutting fluids, their types and applications, minimum quantity lubrication (MQL), dry machining, and application of rotary tools in machining difficult-to-cut materials have been presented in this chapter.

References

1. Neely JE. Basic machine tool operations. Prentice Hall; 2000.
2. Stephenson DA, Agapiou JS. Metal cutting theory and practice. CRC Press; 2016.
3. Trent EM, Wright PK. Metal cutting. Butterworth-Heinemann; 2000.
4. Klocke F, Kuchle A. Manufacturing processes 1: cutting. Berlin, Heidelberg: Springer; 2011.
5. Laperrire L, Reinhart G. Cirp encyclopedia of production engineering. Springer Publishing Company, Incorporated; 2014.
6. El Baradie M. Cutting fluids: part I characterisation. J Mater Process Technol. 1996;56(1–4):786–97.
7. Lawal S, Choudhury I, Nukman Y. Application of vegetable oil-based metalworking fluids in machining ferrous metals—a review. Int J Mach Tools Manuf. 2012;52(1):1–12.
8. Srikant R, Rao D, Subrahmanyam M, Krishna VP. Applicability of cutting fluids with nanoparticle inclusion as coolants in machining. Proc Inst Mech Eng Part J: J Eng Tribol. 2009;223(2):221–5.
9. Daungthongsuk W, Wongwises S. A critical review of convective heat transfer of nanofluids. Renew Sustain Energy Rev. 2007;11(5):797–817.
10. Ghadimi A, Saidur R, Metselaar H. A review of nanofluid stability properties and characterization in stationary conditions. Int J Heat Mass Transf. 2011;54(17):4051–68.
11. Yu W, Choi S. The role of interfacial layers in the enhanced thermal conductivity of nanofluids: a renovated Maxwell model. J Nanopart Res. 2003;5(1–2):167–71.
12. Elcock D. Potential impacts of nanotechnology on energy transmission applications and needs. Argonne National Laboratory (ANL); 2007.
13. Saidur R, Leong K, Mohammad H. A review on applications and challenges of nanofluids. Renew Sustain Energy Rev. 2011;15(3):1646–68.
14. Bhogare RA, Kothawale B. A review on applications and challenges of nanofluids as coolant in automobile radiator. Int J Sci Res Publ. 2013;3(8).

15. Das SK, Choi SU, Yu W, Pradeep T. Nanofluids: science and technology. Wiley; 2007.
16. Swanson EJ, Tavares J, Coulombe S. Improved dual-plasma process for the synthesis of coated or functionalized metal nanoparticles. IEEE Trans Plasma Sci. 2008;36(4):886–7.
17. Li Y, Tung S, Schneider E, Xi S. A review on development of nanofluid preparation and characterization. Powder Technol. 2009;196(2):89–101.
18. Wang X-Q, Mujumdar AS. Heat transfer characteristics of nanofluids: a review. Int J Therm Sci. 2007;46(1):1–19.
19. Sharma P, Sidhu BS, Sharma J. Investigation of effects of nanofluids on turning of Aisi D2 steel using minimum quantity lubrication. J Clean Prod. 2015;108:72–9.
20. Rajmohan T, Sathishkumar S, Palanikumar K. Experimental investigation of machining parameters during turning of Aisi 316l stainless steel using nano cutting environment. In: Applied mechanics and materials. Trans Tech Publications; 2015.
21. Krajnik P, Pusavec F, Rashid A. Nanofluids: properties, applications and sustainability aspects in materials processing technologies. In: Advances in sustainable manufacturing. Springer; 2011. p. 107–13.
22. Choi S, Zhang Z, Yu W, Lockwood F, Grulke E. Anomalous thermal conductivity enhancement in nanotube suspensions. Appl Phys Lett. 2001;79(14):2252–4.
23. Lee C-G, Hwang Y-J, Choi Y-M, Lee J-K, Choi C, Oh J-M. A study on the tribological characteristics of graphite nano lubricants. Int J Precis Eng Manuf. 2009;10(1):85–90.
24. He Y, Jin Y, Chen H, Ding Y, Cang D, Lu H. Heat transfer and flow behaviour of aqueous suspensions of TiO_2 nanoparticles (nanofluids) flowing upward through a vertical pipe. Int J Heat Mass Transf. 2007;50(11):2272–81.
25. Eastman JA, Choi S, Li S, Yu W, Thompson L. Anomalously increased effective thermal conductivities of ethylene glycol-based nanofluids containing copper nanoparticles. Appl Phys Lett. 2001;78(6):718–20.
26. Lee J, Mudawar I. Assessment of the effectiveness of nanofluids for single-phase and two-phase heat transfer in micro-channels. Int J Heat Mass Transf. 2007;50(3):452–63.
27. Pantzali M, Mouza A, Paras S. Investigating the efficacy of nanofluids as coolants in plate heat exchangers (Phe). Chem Eng Sci. 2009;64(14):3290–300.
28. Ding G, Peng H, Jiang W, Gao Y. The migration characteristics of nanoparticles in the pool boiling process of nanorefrigerant and nanorefrigerant-oil mixture. Int J Refrig. 2009;32 (1):114–23.
29. Debnath S, Reddy MM, Yi QS. Environmental friendly cutting fluids and cooling techniques in machining: a review. J Clean Prod. 2014;83:33–47.
30. Mizuhara K, Usui E. Experimental evaluation of cutting fluids penetration. Tribologia (Finland). 1992;11(2):20–9.
31. Machado A, Wallbank J. The effect of extremely low lubricant volumes in machining. Wear. 1997;210(1–2):76–82.
32. Machado A, Wallbank J. The effects of a high-pressure coolant jet on machining. Proc Inst Mech Eng Part B: J Eng Manuf. 1994;208(1):29–38.
33. Kovacevic R, Cherukuthota C, Mazurkiewicz M. High pressure waterjet cooling/lubrication to improve machining efficiency in milling. Int J Mach Tools Manuf. 1995;35(10):1459–73.
34. Pigott R, Colwell A. Hi-jet system for increasing tool life. SAE Technical Paper; 1952.
35. Mazurkiewicz M, Kubala Z, Chow J, Metal machining with high-pressure water-jet cooling assistance—a new possibility. J Eng Ind (Trans ASME). 1989;111(1):7–12.
36. Iyengar HR, Salmon R, Rice W. Some effects of cutting fluids on chip formation in metal cutting. J Eng Ind. 1965;87(1):36–8.
37. Sharma CS, Rice W, Salmon R. Some effects of injecting cutting fluids directly into the chip-tool interface. J Eng Ind. 1971;93(2):441–4.
38. Wertheim R, Rotberg J, Ber A. Influence of high-pressure flushing through the rake face of the cutting tool. CIRP Ann Manuf Technol. 1992;41(1):101–6.
39. Lindeke R, Schoenig F Jr, Khan A, Haddad J. Machining of a, B-titanium with ultra-high pressure through the insert lubrication/cooling. Trans North Am Manuf Res Inst SME. 1991;19:154–61.

40. Thepsonthi T, Hamdi M, Mitsui K. Investigation into minimal-cutting-fluid application in high-speed milling of hardened steel using carbide mills. Int J Mach Tools Manuf. 2009;49(2):156–62.
41. Kipling M. Health hazards from cutting fluids. Tribol Int. 1977;10(1):41–6.
42. Hill E. Microbial aspects of health hazards from water based metal working fluids. Tribol Int. 1983;16(3):136–40.
43. Drozda TJ. Tool and manufacturing engineers handbook: machining, vol. 1. Society of Manufacturing Engineers; 1983.
44. Safety O, Administration H. Metalworking fluids: safety and health best practices manual. US Department Labor; 2001.
45. Foltz GJ. 11 metalworking fluid management and troubleshooting. Metalworking Fluids. 2006; p. 253.
46. Kennedy SM, Greaves LA, Kriebel D, Eisen EA, Smith TJ, Woskie SR. Acute pulmonary responses among automobile workers exposed to aerosols of machining fluids. Am J Ind Med. 1989;15(6):627–41.
47. Hendy M, Beattie B, Burge P. Occupational asthma due to an emulsified oil mist. Occup Environ Med. 1985;42(1):51–4.
48. Bennett E, Bennett D. Occupational airway diseases in the metalworking industry part 1: respiratory infections, pneumonia, chronic bronchitis and emphysema. Tribol Int. 1985;18(3):169–76.
49. Bernstein DI, Lummus ZL, Santilli G, Siskosky J, Bernstein IL. Machine operator's lung: a hypersensitivity pneumonitis disorder associated with exposure to metalworking fluid aerosols. Chest. 1995;108(3):636–41.
50. Suliman S, Abubakr M, Mirghani E. Microbial contamination of cutting fluids and associated hazards. Tribol Int. 1997;30(10):753–7.
51. Hill E, Al-Zubaidy T. Some health aspects of infections in oil and emulsions. Tribol Int. 1979;12(4):161–4.
52. Elsmore R. The survival of *Legionella pneumophila* in dilute metalworking fluids. Tribol Int. 1989;22(3):213–7.
53. Kachan V. Preventing microbial deterioration of cutting fluids. Sov Eng Res. 1987;7(6):68–70.
54. Lloyd G, Lloyd G, Schofield J. Enteric bacteria in cutting oil emulsion. Tribol Int. 1975;8(1):27–9.
55. Rossmoore H, Williams B. Survival of coagulase-positive *Staphylococci* in soluble cutting oils. Health Lab Sci. 1967;4(3):160–5.
56. Pivnick H, Engelhard W, Thompson T. The growth of pathogenic bacteria in soluble oil emulsions. Appl Microbiol. 1954;2(3):140.
57. Shokrani A, Dhokia V, Newman ST. Environmentally conscious machining of difficult-to-machine materials with regard to cutting fluids. Int J Mach Tools Manuf. 2012;57:83–101.
58. Raynor PC, Cooper S, Leith D. Evaporation of polydisperse multicomponent oil droplets. Am Ind Hyg Assoc J. 1996;57(12):1128–36.
59. Hong SY, Broomer M. Economical and ecological cryogenic machining of Aisi 304 austenitic stainless steel. Clean Technol Environ Policy. 2000;2(3):157–66.
60. Sutherland J, Kulur V, King N, Von Turkovich B. An experimental investigation of air quality in wet and dry turning. CIRP Ann Manuf Technol. 2000;49(1):61–4.
61. Dixit US, Sarma D, Davim JP. Environmentally friendly machining. Springer Science & Business Media; 2012.
62. Dornfeld D. Green manufacturing: fundamentals and applications. Springer Science & Business Media; 2012.
63. Sharma VS, Dogra M, Suri N. Cooling techniques for improved productivity in turning. Int J Mach Tools Manuf. 2009;49(6):435–53.
64. Weinert K, Inasaki I, Sutherland J, Wakabayashi T. Dry machining and minimum quantity lubrication. CIRP Ann Manuf Technol. 2004;53(2):511–37.

65. Kalpakjian S, Schmid SR. Manufacturing engineering and technology. Upper Saddle River, NJ, USA: Pearson; 2014.
66. Gupta K, Laubscher R, Davim JP, Jain N. Recent developments in sustainable manufacturing of gears: a review. J Clean Prod. 2016;112:3320–30.
67. Canter N. The possibilities and limitations of dry machining. Tribol Lubr Technol. 2009;65 (3):40.
68. Klocke F, Eisenblätter G. Dry cutting. CIRP Ann Manuf Technol. 1997;46(2):519–26.
69. Kishawy H, Becze C, McIntosh D. Tool performance and attainable surface quality during the machining of aerospace alloys using self-propelled rotary tools. J Mater Process Technol. 2004;152(3):266–71.
70. Kishawy H, Wilcox J. Tool wear and chip formation during hard turning with self-propelled rotary tools. Int J Mach Tools Manuf. 2003;43(4):433–9.
71. Kishawy H, Shawky A, Elbestawi M. Assessment of self-propelled rotating tools during high speed milling. In: SME proceedings of the fourth international machining & grinding conference, Troy Michigan, USA. 2001.
72. Shaw M. The rotary cutting tool. Trans ASME. 1952;74:1065.
73. Venuvinod P, Rubenstein C. The principle of equivalent obliquity and its application to rotary cutting. CIRP Ann Manuf Technol. 1983;32(1):53–8.
74. Armarego E, Karri V, Smith A. Fundamental studies of driven and self-propelled rotary tool cutting processes—I. Theoretical investigation. Int J Mach Tools Manuf. 1994;34(6):785–801.
75. Armarego E, Karri V, Smith A. Fundamental studies of driven and self-propelled rotary tool cutting processes—II. Experimental investigation. Int J Mach Tools Manuf. 1994;34(6):803–15.
76. Armarego E, Katta R. Predictive cutting model for forces and power in self-propelled rotary tool turning operations. CIRP Ann Manuf Technol. 1997;46(1):19–24.
77. Chang X, Chen W, Pang X, Zhong G. Selection of cutting regime for self-propelled rotary cutting tools. Proc Inst Mech Eng Part B: J Eng Manuf. 1995;209(1):63–6.
78. Kishawy HA, Gerber AG. A model for the tool temperature during machining with a rotary tool. In International mechanical engineering congress and exposition symposium on fundamental issues in machining. 2001.
79. Kishawy H, Pang L, Balazinski M. Modeling of tool wear during hard turning with self-propelled rotary tools. Int J Mech Sci. 2011;53(11):1015–21.
80. Luo S, Liao Y, Tsai Y. Wear characteristics in turning high hardness alloy steel by ceramic and Cbn tools. J Mater Process Technol. 1999;88(1):114–21.

Index

H. A. Kishawy and A. Hosseini, *Machining Difficult-to-Cut Materials*, Materials Forming, Machining and Tribology, https://doi.org/10.1007/978-3-319-95966-5

The manufacturer's authorised representative in the EU is Springer
Nature Customer Service Centre GmbH, Europaplatz 3, 69115 Heidelberg,
Germany. If you have any concerns regarding our products, please
contact ProductSafety@springernature.com

Printed and bound by CPI Group (UK) Ltd, Croydon, CR0 4YY
05/10/2026
02228138-0008